Diagnosing Male Infertility

Progress in
Reproductive Biology
and Medicine

Vol. 15

Series Editor
M. L'Hermite, Brussels

KARGER

Basel · München · Paris · London · New York · New Delhi · Bangkok · Singapore · Tokyo · Sydney

Diagnosing Male Infertility

New Possibilities and Limits

Volume Editors
Giovanni M. Colpi, Milan
Diego Pozza, Rome

51 figures and 67 tables, 1992

Basel · München · Paris · London · New York · New Delhi · Bangkok · Singapore · Tokyo · Sydney

Progress in Reproductive Biology and Medicine

Library of Congress Cataloging-in-Publication Data
Diagnosing male infertility: new possibilities and limits / volume editors, Giovanni M. Colpi,
Diego Pozza. (Progress in reproductive biology and medicine; vol. 15)
Includes bibliographical references and index.   (alk. paper)
1. Infertility, Male – Diagnosis.   I. Colpi, Giovanni M.   II. Pozza, Diego   III. Series.
[DNLM: 1. Infertility, Male – diagnosis.   W1 PR681BD v. 15 / WJ 709 D5355]
RC889.D49   1992   616.6′92 – dc20
ISBN 3–8055–5443–5

Drug Dosage
The authors and the publisher have exerted every effort to ensure that drug selection and dosage set forth in this text are in accord with current recommendations and practice at the time of publication. However, in view of ongoing research, changes in government regulations, and the constant flow of information relating to drug therapy and drug reactions, the reader is urged to check the package insert for each drug for any change in indications and dosage and for added warnings and precautions. This is particularly important when the recommended agent is a new and/or infrequently employed drug.

Contents

Preface

At present the medical responses to the increased incidence of infertile couples have become highly technical, with recourse to ever more sophisticated methods for assisted procreation. These are costly, economically and psychologically, to the couple and to the general population in terms of public health costs. For these reasons, the role of the andrologist has become one of increasing importance.

A generation ago, it was almost exclusively the woman of the infertile couple who was studied in depth. In the 1970s, specialists began to pay attention to the male partner. However, now that assisted procreation has become so widely practiced, it often occurs that the only male component used for study is the spermatozoon.

It sometimes appears to be forgotten that the success of any assisted procreation method also depends on the overall quality of the semen sample. The fertilizing ability of spermatozoa from severely abnormal semen (because of a varicocele, for one example) is very probably less than that of spermatozoa from an improved semen sample from the same patient after surgery.

In addition, we have seen in our clinical practices that appropriate treatment (surgery, microsurgery, hormonal, etc.) of severely infertile patients has sometimes resulted in pregnancies after natural intercourse or simple insemination for couples who have been told at other centers to seek assisted procreation, or perhaps donor sperm insemination, or have even tried these without success. We feel that all of this emphasizes how important it is for clinical, biological and ethical reasons that modern infertility centers include an andrologist to work in close cooperation with the gynecological team.

We sincerely hope that in the near future methods will become available for diagnosing more and more precisely the causes or associated factors in semen anomalies, whether they be secretory, excretory or combinations of these. Without these, it is difficult to apply the most effective medical, surgical or biological treatment for improving the overall quality of the semen. We can hope to have tests that will show with high probability whether or not the spermatozoa in a given ejaculate will be able to fertilize and thus avoid using in vitro fertilization as a diagnostic procedure.

We have not designed this volume to be an organized presentation of everything known at present about male infertility. It is a collection of papers describing new studies in the field of diagnosis of male infertility. We hope it will bring the andrologist up to date and stimulate more clinical investigations.

January 1992   *Giovanni M. Colpi · Diego Pozza*

Colpi GM, Pozza D (eds): Diagnosing Male Infertility.
Prog Reprod Biol Med. Basel, Karger, 1992, vol 15, pp 1–6

Male Infertility Now and in the 1990s

Alpay Kelâmi
Berlin, FRG

Before trying to speculate on the future of male infertility, I think it would be worthwhile to try to explain what andrology is, or even better, what I understand under *andrology.*

Andrology is the exact counterpart to gynecology, and thus deals with *fertility and sterility* as well as *sexual function and dysfunction* of the *male.* Anatomically starting at the scrotum and testis, extending towards the urethra and ending at the meatus or preputial opening, andrological organs form an independent system (fig. 1). The andrological and urological systems are connected where the ejaculatory ducts join the verumontanum. The urethra serves as a conduit for both urine and ejaculate. Hypospadias causes not only urological problems but, due to penile deviation, andrological problems as well. Disorders of the prostate and subsequent operations can have andrological consequences such as infertility due to failure of prostate secretion or retrograde ejaculation. Testicular tumors have purely andrological consequences, unless lymph node metastases cause deviation of the ureter.

Andrology is not only the science of fertility and sterility. Disorders that render coitus difficult or impossible, such as congenital or acquired penile deviations and erectile dysfunction, have been added to it in the last two decades [Kelâmi, A.: Atlas of Operative Andrology, de Gruyter, Berlin, 1980]. In spite of the fantastic developments in the field of the female reproductive system with all the new artificial possibilities, most children are still being conceived in a way that is not manipulated by medicine; by the act of copulation. Frequently, I try to explain andrology as some creature walking on two legs, one being fertility-sterility, and the other being sexual function-dysfunction (fig. 2).

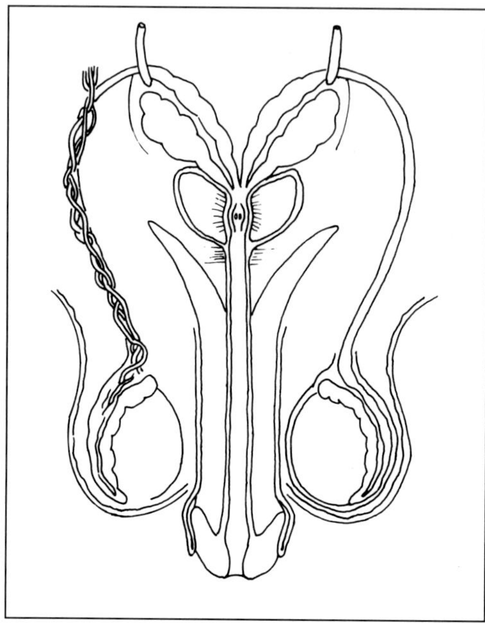

Fig. 1. Andrological organ system.

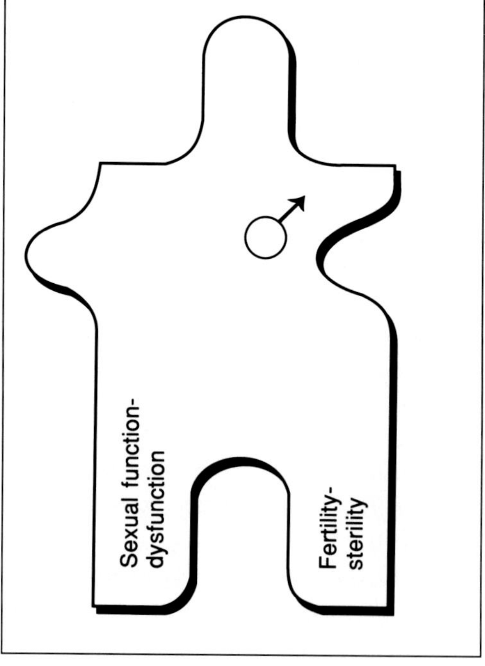

Fig. 2. Andrology man on two legs: sexual function-dysfunction; fertility-sterility.

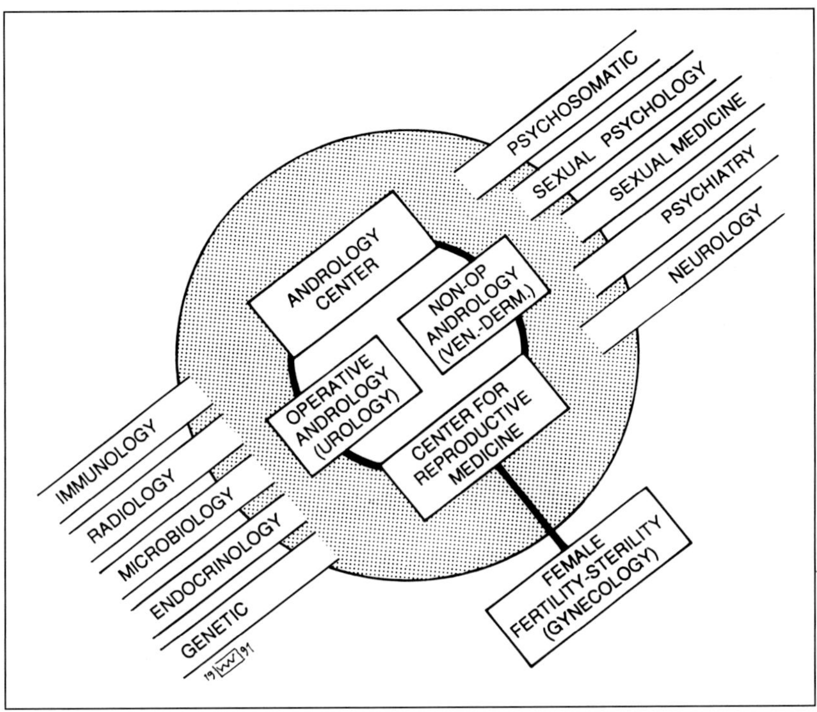

Fig. 3. Andrology: a multidisciplinary specialty.

What is an andrologist? What can an andrologist do? What should an andrologist do? An andrologist is a medical doctor who deals with male genital organs. What an andrologist can do depends very much on his background, on the specialty from which the andrologist comes. This varies again very much from country to country. An andrologist can be, for example, a dermato-venerologist, a urologist, a gynecologist, a general surgeon, an endocrinologist or a general internist, etc. Figure 3 shows the situation in which andrology should be practiced. The inner ring is typical for Germany. The outer ring can be extended as needed and as newer disciplines develop. In today's Germany, there are doctors who practice andrology without being able to call themselves andrologists. It is not yet a recognized specialty. Each country has its own specific traditional and/or political version. The decisive point is that andrology today can only be practiced as a multidisciplinary specialty (fig. 3).

Table 1

Parameters of male fertility	ejaculate: volume, viscosity, pH sperm: number, motility, morphology fructose, immunology
Hormonal factors	hypothalamus, GnRh pituitary, LH, FSH testis (Leydig cells, tubuli) hyperprolactinemia
Primary disorders of germinal cells	tubular insufficiency Sertoli-cell-only syndrome orchitis, tumors
Genetic factors	chromosomal abnormalities (Klinefelter syndrome, XXY) defects of spermatozoa (globozoospermia, immotile cilia syndrome) hereditary-degenerative syndromes (mucoviscidosis)
Vascular factors	varicocele
Dysfunction of epididymis, prostate and seminal vesicles	infection, stenosis, obstruction, Young syndrome
Embryological factors	maldescensus testis, anorchia
Congenital factors	aplasia of vas deferens aplasia of epididymis
Obstructive factors	infection of vas deferens, epididymis, ejaculatory ducts azoospermia
Ejaculatory factors	ejaculatio praecox retrograde ejaculation urethral strictures neurologic disorders (MS)
External factors	alcohol, nicotine, drugs, environmental factors (radiation, heat) iatrogenic-surgical (herniotomy, orchidopexy, pelvic surgery)

After its establishment as a specialty, it will certainly be easier to train specialists at certain andrology centers. In order to make andrology better known to the public and to give the andrologically interested doctors an opportunity to discuss andrological problems, series of different meetings are being organized:

'Berliner Andrological Colloquium', Berlin; Kelâmi, A., Nürnberger, F.
'International Symposon of Operative Andrology', Berlin and London;
   Kelâmi, A., Pryor, J.P.
'Art and Andrology 1–5', Berlin; Kelâmi, A.
'International Meeting of Andrology', Rome and Como; Pozza, D., Colpi, G.M.
'International Forum of Andrology', Paris; Arvis, G.

Concerning the future of andrology, I will try to outline some fertility-sterility problems, since these are the topics of the present meeting. Diagnostic methods, parameters, and factors which influence fertility are numerous (table 1).

We are far from understanding the interrelations between these entities and how to influence them. Why can a couple not have a child, even though all parameters are perfect? One man with varicocele can fertilize an egg, the other cannot. The number of sperm one needs for fertilization is growing smaller and smaller every year. Why do we need so and so many millions/milliliter when just one would be enough? This list could be extended indefinitely. In the nonoperative sector, treatment has mostly been empiric, the results not always reproducible and not very encouraging. I do not think there will be any great breakthrough in this area in the next decade.

In the operative sector, some improvements have been made. Microsurgical techniques, epididymal aspiration, alloplastic spermatocele, and combination with the gynecological insemination program (IVF, ET, GIFT, ZIFT) promise better results. However, one still should not expect a great breakthrough. Greater success will probably be obtained in the hands of a few andrological surgeons in andrological centers, at least in the beginnig. It makes a big difference if surgery is done daily, weekly, or only a few times a year. Compared with gynecology as our counterpart, andrology is a rather new branch of medicine. Therefore, we need more time, more established institutions, more basic and clinical research, more recognition, and more respect. It is not enough to have just one or two names in each country or in a city (some do not have any at all).

The problems are present; the patients bring the problems with them. But if we are honest, we cannot promise them too much, and I do not think this will change in the very near future.

It is sometimes a hard decision, but a very important and correct decision to tell the patients the truth – the truth that he cannot have his own child – and at the same time to offer him the alternatives of: (1) heterologous insemination; (2) adoption, or (3) giving up the wish to have a child. The happiness of a couple is also possible if the truth is known, well explained, and understood.

All in all, I do expect a little improvement, simply because the number of andrologists and scientific work being done are increasing worldwide. Let us be realistic, and at the same time optimistic and work for a better future of andrology.

Prof. Dr. Alpay Kelâmi, Brahmsstrasse 32, D-W–1000 Berlin 45 (FRG)

Colpi GM, Pozza D (eds): Diagnosing Male Infertility.
Prog Reprod Biol Med. Basel, Karger, 1992, vol 15, pp 7–18

Semen Analyses in the Nineties: Some Reflections on Their Context, Importance and Limits

Marco Balerna, Ana Piffaretti-Yanez[1]

Andrology Laboratory, Endocrinological Gynecology Unit, 'La Carità Hospital',
Locarno, Switzerland

Semen analyses (SAs) still play a central role in the diagnosis, therapy follow-up and fertility prognosis of male/couple infertility. In approximately 90% of the cases SAs may be used to determine the cause of male infertility (table 1). Although SA is often considered as a simple routine laboratory analysis, it involves various procedures of a special difficulty grade, requiring skill, mental thinking and conscientious work. Herein we will review some basic points and discuss recent advances made in this area.

Definition

SA is nowadays an overall term encompassing a whole series of cytological, biological, microbiological, biochemical and immunological methods performed on the cellular and fluid components of fresh semen samples.

From a technical point of view, SAs may be subdivided into three different levels of analyses: (1) assessment of sperm number, subjective motility, morphology; simple functional and immunological tests; (2) seminal biochemistry and microbiology, and (3) ultrastructural cell biology; cellular biochemistry and immunology.

[1] The authors wish to thank the Swiss National Research Foundation for partial financial support (Grant No. 3.937-0.87 to M.B.) and the whole staff of our laboratories for their invaluable and dedicated work.

Table 1. Causes of male infertility [reproduced from 50]

		Frequency, %
Untreatable sterility	Primary testicular failure	12
Potentially treatable conditions	Secondary testicular failure Genital tract obstruction Sperm autoimmunity Coital disorders Toxin effects Epididymal necrospermia	13
Untreatable subfertility	Oligozoospermia Asthenozoospermia Teratozoospermia	75

At present, SAs are seldom, if at all, considered as 'sperm functional (bio)assays' but are often integrated with such tests (see below: SAs and the So-Called 'Functional Tests').

Operative and Physiological Prerequisites

To be correctly performed and interpreted, SAs need that a whole series of conditions are respected and followed.

Clinical Conditions

The clinician must inform his patients about the necessity to provide a 'correct and complete' semen sample and must explain to them how this can be accomplished [1, 2]. Factors which may influence the ejaculate composition (such as psychologic stress, sexual activity, effect of heat on testicular processes, occupational hazards, season, etc.) should be known by the clinician and explained to the patient. In particular the patient should be informed about: (a) the necessity of a sexual abstinence period; (b) the advantages and disadvantages of the various acceptable methods for semen collection; (6) the hygienical conditions to observe during sample collection, and (d) the problems encountered in obtaining the ejaculate in or outside the clinic/hospital. In the latter case the patient must be told how to properly transport the sample and to notify the laboratory how much time had elapsed from ejaculation. The use of special sperm-collecting devices should also be considered [3].

Technical Information and Training of Biologists and Technicians Performing SAs

(1) Since ejaculation is not homeostatically controlled, the analysts should recognize the 'uniqueness' of each semen sample as reflecting the andrological (testicular, epididymal and male accessory glands) state of a man at a given moment of his sexual life. Furthermore, there is still controversy concerning the stability of basic semen parameters as function of time [2, 4, 5, 16]. The biologists and technicians should, therefore, consider a semen specimen in a very different manner than other types of body fluids.

(2) The lack of homeostasis in the ejaculation process explains, at the same time, why a single SA can only partially characterize the quality of semen produced by a single man. The principle of a 'trend toward the mean' of the semen variables as function of time for a given man is an important point in this context [6, 7].

(3) Furthermore, evidence suggesting that each semen specimen is composed of various sperm subpopulations having different degrees of cellular maturation has been presented but only recently is achieving recognition [see e.g. 8, 9]. Ejaculates collected subsequently or after different abstinence periods could therefore be thought to be composed of different sperm subclasses. This finding could partly explain the difficulties encountered in using SAs data for fertility prognostic purposes, and constitute a further physiologic explanation for semen variability.

(4) Finally, a particular training period for biologists/technicians in a reference laboratory is of crucial importance so that they can properly understand and organize the analytical work in their own laboratory [10, 11].

Technological Aspects

(1) Semen is only seldom 'sterile'. In order to be able to analyze the original seminal flora without introducing external contaminants during collection, ejaculates should be obtained in sterile, wide-mouth containers that can be easily and tightly capped by the patient immediately after ejaculation. Another aspect that must be considered is the material of which the container is made, since it may influence the viability of spermatozoa and seminal plasma characteristics [12–14].

(2) The laboratory needs to know if the semen sample produced by the patient is complete or incomplete (semen loss *during* or *after* collection?).

(3) Upon receiving the ejaculate, the technician *must* immediately induce contact between the prostatic and the vesicular components by vortexing the sample. Moreover, all samples should be incubated at the same temperature and if possible at 37 °C.

(4) The laboratory personnel must also know what parameters are to be measured, what procedures are to be followed and with which priority the various assessments will be made (e.g. aliquots needed for microbiology and for pH measurements should be drawn immediately after sperm liquefaction from the capped container after well mixing the sample again).

(5) The analytical methods used by the laboratory should be the most sensitive, specific, precise, repeatable and cost-effective as possible (e.g. acid phosphatase assays are being replaced by the more cost-effective zinc assays; this also solves the problem connected to the instability of the enzyme).

(6) The laboratory needs both 'internal' and 'external' control protocols [11]. If possible, it should establish its own 'normal values' and check at least some of its assessments with other laboratories ('inter-laboratory cross-check' of, e.g. sperm number, morphology and seminal plasma biochemistry). To assess the specificity and sensibility of the laboratory's results and also to ascertain whether a given semen variable has a significant diagnostic and/or prognostic value, the use of simple but very powerful statistical methods should be reminded here (odds ratios, likelihood ratios) [15].

Static and Functional Character of SAs

As any other clinical laboratory procedure, SAs have an intrinsic *static* character: one only obtains from them numerical or categorical indications of the measured variables. SAs results become *functional* when interpreted in the wider context of patient's history, physical examination and complementary laboratory assessments (e.g. endocrinological and genetic analyses) [51].

SAs in the Diagnosis and Therapy of Male Infertility

Many efforts have been made in recent years to better understand and optimize the diagnostic power of SAs. The work done under the auspices of WHO was – and still is – of particular importance in this context [16–19].

Table 2. Influence of WHO normal values in the classification of men consulting for infertility[1]

	Criteria WHO normal values	Cases	
		n	% of total
Total number of analyses		6,831	100.0
Number of cases remaining after elimination of those with:			
1 Sperm count	≤ 20 million/ml	841	12.3
2 Sexual abstinence	< 2 or > 7 days	733	10.7
3 Volume	≤ 2 ml	664	9.7
4 pH	< 7.2 or > 7.8	599	8.8
5 Sperm count	≥ 250 million/ml	597	8.7
6 Morphology	normal form $\leq 50\%$	526	7.7
7 Motility	progressive linear and progressive nonlinear motility $\leq 50\%$	461	6.7
8 WBC	≥ 1 million/ml	458	6.7
Total number of ejaculates with all-normal semen parameters		458	6.7

A total of 6,831 SAs were reviewed and classified using the normal values issued by WHO (1987). Cases were eliminated in a sequential manner following the order and criteria shown in the table. Only 458 men (6.7% of the total) were found to have all semen parameters normal according to the WHO guidelines. As has been previously criticized, the normal values of WHO tend to place the majority (93%) of the patients in the potentially infertile or subfertile group [15, 20, 21, 30, 31].

[1] Semen analyses performed in the Andrology Laboratory of the Endocrinological Gynecology Unit, 'La Carità' Hospital, Locarno (August 1978 to December 1987).

However, the 'normal values' issued by WHO (1987) are still object of research and some criticism [20, 21] (table 2).

Diagnostically, SAs are important in unraveling the factor(s) causing the male's and eventually also the couple's infertility. In particular, they are useful to detect: (a) an insufficient sperm production in the testes (low sperm number in the absence of obstructions but with sufficient and active testicular tissue); (b) occlusions at some level of the male genital tract and/or (partial) retrograde ejaculation; (c) defects of the coagulation/liquefaction mechanisms; (d) an insufficient or abnormal prostatic and/or vesicular secretion; (e) the presence of infections and/or inflammations of the male genital tract (epididymitis, vesiculitis, prostatitis and combined patterns accompanied or not by leukospermia); (f) the presence of (auto)antibodies to sperm in the seminal plasma or at the sperm membrane level, and (g) the presence of bacteria, Ureaplasma, Mycoplasma, Chlamydia or viruses which would eventually be associated with sperm (hazards in assisted reproduction procedures and for sperm cryoconservation).

SAs are also useful to assess disturbances of sperm motility and epididymal maturation of gametes. Furthermore, techniques used in SAs form the basic criteria on which to judge the success of vasectomy/varicocelectomy and the goodness of the gamete preparations for assisted reproductive techniques (IVF, GIFT, ZIFT, etc.). Finally, besides their importance in basic cell biological, biochemical and epidemiological research, SAs are also useful to detect toxic factors in semen thought to impair testicular function such as heavy metal ions, pesticides, etc.

The usefulness of SAs in diagnosis and therapy is however limited by several potential biases, hazards and technical limits:

(a) Manipulation of semen implies a series of hazards which should be known and prevented by all operators involved [22].

(b) It is often forgotten that SAs are performed under 'nonphysiologic' conditions (seminal plasma has repeatedly been demonstrated to be potentially 'dangerous' – e.g. motility impairing – for spermatozoa [23–25]).

(c) The assessments performed during SAs are difficult to be made quantitative. Even the simplest technical steps are coupled to a series of (sometimes unrealized) methodological biases that are difficult to control or prevent [26]. The efforts made until now to incorporate both hard- and software in overcoming many of the problems presently encountered at the analytical level must still be qualified as 'not yet optimal' [38]. It can however be forecasted that in the near future we will be provided with powerful – eventually robotic-type – hardware which may resolve the old 'number, motility and morphology problem'. Other simple methods to standardize laboratory observations exist [27], but are not well known or then, surprisingly, not applied.

(d) On the immunological level, only a few 'basic' tests to detect (auto)antisperm antibodies were agreed upon at the international level. The need to elucidate the many, and often hidden, immunological aspects of male and induced female infertility is still one of the most controversial aspects of research in spermatology.

All these points – and others which will be addressed in the next section – diminish and severely restrict the power of the studies aiming at the evaluation of surgical, pharmacological and endocrinological treatments of male infertility. A well-organized clinical study should therefore not only adopt correct experimental designs and modern statistical evaluation

Table 3. Statistical methods used in recent years for the set-up of studies and data analysis in diagnosis, therapy and fertility prognostic studies based on semen analyses

---

A Basic concepts and study planning methodology[1] [28, 29, 42, 43, 49][2]

B Data analysis[1]
   1 Parametric vs. nonparametric data analyses
      Transformation ('normalization') of data sets [17, 36, 47]
   2 Life-tables method [42, 45]
   3 Couple-month method [35]
   4 Kaplan-Meyer survival curves
   5 Cox's proportional hazards (multiple) regression model [30, 46]
   6 Receiver operating characteristic (ROC) curves [31, 40, 41]
   7 Logistic regression
   8 Recursive partitioning analysis
   9 Specificity, sensibility, odds ratios, likelihood ratio rates, pre- and posttest probability calculations [15]

---

[1] Numbers in brackets refer to references listed herein. No specific reference is given for general methods.
[2] The 'pater semper incertus' principle is often not taken into account.

---

methods (table 3), but should also take into account and minimize all the possible and forementioned biases [28, 29].

The Prognostic Value of SAs

The various and complex phenomena taking place at ejaculation clearly indicate why only in a very reduced number of cases the fertility of a given man can fairly be predicted by SAs (e.g. aspermia, azoospermia [for a former bibliography, see Schirren [1]). The use of so-called 'cut-off values' of seminal variables able to discriminate fertile from subfertile or infertile men has again received considerable criticism [15]. It is particularly unfortunate that, even at the present moment, different workers make contrasting statements about the fertility-predicting power of seminal variables [cf. 20, 21, 30, 31]. Although the dispute is far from being concluded, two different factors acting concomitantly are making the discussion of this issue somewhat less important. First, the data from former retrospective studies are now being strongly questioned on the basis of the inherent difficulties and potential biases often associated with this type of study (table 3, A), and because of the inability to ascertain by effective means that the wives of the infertile couples were truly fertile. Second, the process of 'short-circuiting' the male factor brought about by the most recent techniques of assisted reproduction, such as zona drilling, subzona fertilization and sperm microinjection procedures [32, 44] also lessens the importance of the above-mentioned 'cut-off values'.

SAs and the So-Called 'Functional Tests'

Since the relationship between conventional seminal parameters and male/couple fertility has increasingly been recognized to be a loose one, several workers have proposed bioassays ('functional tests') claiming that these tests were able to forecast the fertility potential of a given man with high sensibility and specificity. The experiences made in the last 10 years or so have amply demonstrated that none of these tests is actually able to do that. In a certain sense, this is an unfortunate situation; however, the diagnostic power of SAs would assuredly gain by being coupled to the prognostic power of some 'fertility potential bioassays'.

As a particular example one could mention here the hypo-osmotic swelling (HOS) test [47, 48]. Devised in 1984 and in spite of early reports on its (good) correlation with the fertilization potential of a given sperm specimen, the HOS test could not yet be demonstrated unequivocally to be a 'fertility test'. We believe that the initial claims were 'rather too easily' accepted by the scientific community and raised hopes which could not be substantiated. This is particularly disturbing, as such claims forced other workers to re-study, re-check and re-demonstrate the contrary of the original claims. Even the 'best' bioassay introduced so far in terms of sensibility but not of specificity [15], i.e. the sperm penetration assay (SPA), is well known to still suffer from a rather high rate of false-negative results [15, 33, 34]. Its clinical value has been proposed to be higher in the case of positive test results.

Thoughts About (and the Future of) SAs

These authors were once told that '... it is very interesting to listen to andrologists' lectures, but at the end what an infertile couple needs is the proper sperm fertilizing the proper oocyte ...!'

Some years ago, Hargreave and Elton [35] asked in a rather provocative way: 'Is conventional sperm analysis of any use?' Recently, Comhaire [19] considered that '… the fact that andrology has not yet been accepted as a full and independent medical speciality has hampered scientific progress in the field of objective assessment of male infertility and its causes'. On the other hand, Eliasson [36] admonishes that: 'I am personally convinced that if andrologists gave more of their time and resources to the study of a man as an individual and as a part of a male population rather than as a potentially fertile partner, we would more rapidly than now be able to collect unbiased information of the value for the man, the society and the infertile couple'.

All these observations reflect well the situation in the field of SAs. The problem is, in fact, that until now even with the advances made in the diagnostic area, it is still difficult for andrologists to advance in an equilibrated manner, coping with both basic and clinical research and at the same time with the practical needs of infertile men and of their practitioners. Since infertile couples and their doctors want primarily to cure the infertility at least temporarily (and nothing else), heavy pressure has compelled the search for means of 'short-circuiting' the male factor. In other words, also the andrologists were, and still are, forced to concentrate on the search for the cure rather than the cause of infertility. The need to achieve pregnancy against all odds has given rise to the impressive, and assuredly not all-justified, explosion of assisted reproduction methods (accompanied by other more basic attacks to some 'central dogma', as Silber's [37] epididymal maturation dilemma).

Andrologists themselves have contributed to the present condition. We must admit that they were not yet able to produce the technological advance(s) which in turn would have induced much more scientific and clinical interest in their discipline. As a consequence, they had to follow rather than lead. We feel therefore that it is important that in the near future andrologists become able (and allowed to be free!) to cope with classical methodologies, basic research novelties, and the newer techniques introduced by assisted reproduction, not dismissing one or the other of these themes. Moreover, if one thinks that entire chapters of SAs (like seminal virology, microbiology, preparative sperm cytology, sperm biochemistry and physiology or then sperm micromanipulation) are either no man's land or then clay ground, it is clear that there is an urgent need for renewed and strong efforts from andrologists [39, 44].

In conclusion, we believe that SAs will maintain their central role in fertility/infertility routine and research work, if andrologists will be able in the future to innovate, refresh and produce further sound work on SAs.

References

1 Schirren C: Assessment of fertility in the human male from semen examinations (315 references). Bibliogr Reprod 1982;40:205.

2 Poland ML, Moghissi KS, Giblin PT, Ager JW, Olson JM: Stability of basic semen measures and abnormal morphology within individuals. J Androl 1986;7:211–214.

3 Zavos PM, Goodpasture JC: Clinical improvements of specific seminal deficiencies via intercourse with a seminal collection device versus masturbation. Fertil Steril 1989;51:190–193.

4 Schwartz D, Laplanche A, Jouannet P, David G: Within-subject variability of human semen in regard to sperm count, volume, total number of spermatozoa and lenght of abstinence. J Reprod Fertil 1979;57:391–395.

5 Heuchel V, Schwartz D, Czyglik F: Between and within subject correlations and variances for certain semen characteristics in fertile men. Andrologia 1982;15:171–176.

6 Baker HWG, Kovacs GT: Spontaneous improvement in semen quality: regression toward the mean. Int J Androl 1985;8:421–426.

7 Bostofte E, Forrest M, Andersen B: Regression toward the mean in fertility research – an empirical study of semen quality. Andrologia 1986;19:76–79.

8 Ben-Nun I, Lancet M, Danon D: Separation of human spermatozoa from a single ejaculate into various age groups. Int J Fertil 1980;25:127–130.

9 Huszar G, Vigue L, Corrales M: Sperm creatine kinase activity in fertile and infertile oligospermic men. J Androl 1990;11:40–46.

10 Schirren C: Differenz in den Labordaten zwischen Praxis- und Klinik-Laboratorien. Andrologia 1986;18:332–334.

11 Dunphy BC, Kay R, Barratt CLR, Cooke IAD: Quality control during the conventional analysis of semen, an essential exercise. J Androl 1989;10:378–385.

12 Balerna M, Nutini L, Eppenberger U, Campana A: Technology and instrumentation for semen analyses and AIH/AID. Effect of plastic and glass on sperm motility, pH and oxidation. Arch Androl 1985;15:225–230.

13 Strikland DM, Ziaya PR: Reduced sperm motility in plastic containers. Lab Med 1987;18:310–312.

14 Check JH, Shanis BS, Wu Ch, Bollendorf A: Evaluating glass, polystyrene and polypropylene containers for semen collection and sperm washing. Arch Androl 1988;20:251–255.

15 Polansky FF, Lamb EJ: Analysis of three laboratory tests used in the evaluation of male fertility: Baye's rule applied to the postcoital test, the in vitro mucus migration test and the zona-free hamster egg test. Fertil Steril 1989;51:215–228.

16 WHO: WHO Laboratory Manual for the Examination of Human Semen and

Semen-Cervical Mucus Interaction. Cambridge, Cambridge University Press, 1987, pp 1–67.

17   WHO (Comhaire/de Kretser/Farley/Rowe): Towards more objectivity in diagnosis and management of male fertility. Int J Androl. 1987;suppl 7.

18   Rowe PJ: WHO's approach to the management of the infertile couple; in Negro-Vilar A, Isidori A, Paulson J, Abdelmassih R, de Castro MPP (eds): Andrology and Human Reproduction. Serono Symposia. New York, Raven Press, 1988, vol 47, pp 291–309.

19   Comhaire F: Evaluation of male infertility; in Serio M (ed): Perspectives in Andrology. Serono Symposia. New York, Raven Press, 1989, vol 53, pp 45–56.

20   Polansky FF, Lamb EJ: Do the results of semen analyses predict future fertility? A survival analysis study. Fertil Steril 1988;49:1059.

21   Dunphy BC, McNeal LM, Cooke ID: The clinical value of conventional semen analysis. Fertil Steril 1989;51:324–329.

22   Schrader SM: Safety guidelines for the andrology laboratory. Fertil Steril 1989;51:387–389.

23   Jones R, Mann Th, Sherins R: Peroxidative breakdown of phospholipids in human spermatozoa, spermicidal properties of fatty acid peroxides and protective action of seminal plasma. Fertil Steril 1979;31:531–537.

24   Aitken RJ, Clarkson JS, Fishel S: Generation of reactive oxygen species, lipid peroxidation and human spermatozoa function. Biol Reprod 1989;41:183–197.

25   Rao B, Soufir JC, Martin M, David G: Lipid peroxidation in human spermatozoa as related to midpiece abnormalities and motility. Gamete Res 1989;24:127–134.

26   Menkveld R, van Zyl JA, Kotze TJvW: A statistical comparison of three methods for the counting of human spermatozoa. Andrologia 1984;16:554–558.

27   Katz DF, Diel L, Overstreet JW: Differences in the movement of morphologically normal and abnormal human seminal spermatozoa. Biol Reprod 1982;26:566–570.

28   Olive DL: Analysis of clinical fertility trials: a methodological review. Fertil Steril 1986;45:157–171.

29   Evers JLH: The pregnancy rate of the no-treatment group in randomized clinical trials of endometriosis therapy. Fertil Steril 1989;52:906–907.

30   Bostofte E: Prognostic parameters in predicting pregnancy. A 20-year follow-up study comprising semen analysis in 765 men of infertile couples evaluated by the Cox regression model. Acta Obstet Gynecol Scand 1987;66:617–624.

31   Hinting A, Comhaire F, Schoonjans F: Capacity of objectively assessed sperm motility characteristics in differentiating between semen of fertile and subfertile men. Fertil Steril 1988;50:635–639.

32   Boldt J: Micromanipulation in human reproductive technology. Fertil Steril 1988;50:213–215.

33   Ausmanas M, Tureck RW, Blasco L, Kopf GS, Ribas J, Mastroianni L: The zona-free hamster egg penetration assay as a prognostic indicator in a human in vitro fertilization program. Fertil Steril 1985;43:433–437.

34   Corson SL, Batzer FR, Marmar J, Maslin G: The human sperm-hamster egg penetration assay: prognostic value. Fertil Steril 1988;49:328–334.

35   Hargreave TB, Elton RA: Is conventional sperm analysis of any use? Br J Urol 1983;55:774–779.

36   Eliasson R: Comments to some basic problems in andrology; in Negro-Vilar A, Isidori A, Paulson J, Abdelmassih R, de Castro MPP (eds): Andrology and Human Reproduction. Serono Symposia. New York, Raven Press, 1988, vol 47, pp 39–44.

37   Silber SJ, Balmaceda J, Borrero C, Ord T, Asch R: Pregnancy with sperm aspiration from the proximal head of the epididymis: A new treatment for congenital absence of the vas deferens. Fertil Steril 1988;50:525–528.

38   Knuth UA, Yeung CH, Neuwinger J, Nieschlag E: Computerized semen analysis; in Serio M, Waites G (eds): Recent Advances in Andrology. Serono Symposia Rev No 20. Rome, Ares-Serono Symposia, 1989, pp 13–26.

39   Acosta A, Swanson RJ, Ackerman SB, Kruger TF, van Zyl JA, Menkveld R: (eds): Human Spermatozoa in Assisted Reproduction. Baltimore, Williams & Wilkins, 1990.

40   Peng H-Q, Collins JA, Wilson EH, Wrixon W: Receiver-operating characteristics curves for semen analysis variables: Methods for evaluating diagnostic tests of male gamete function. Gemete Res 1987;17:229–236.

41   DeLong ER, DeLong DM, Clarke-Pearson DL: Comparing the areas under two or more correlated receiver operating characteristic curves: a nonparametric approach. Biometrics 1988;44:837–845.

42   Tulandi T, Cherry N: Clinical trials in reproductive surgery: randomization and life-table analysis. Fertil Steril 1989;52:12–14.

43   Wheeler JM: The emperor (or rather, his statistician) has new clothes. Fertil Steril 1990;53:220–222.

44   Ng S-C, Bongso A, Sathananthan H, Ratnam SS: Micromanipulation; its relevance to human in vitro fertilization. Fertil Steril 1990;53:203–219.

45   Schwartz D, Mayaux MJ: Mode of evaluation of results in artificial insemination; in David G, Price WS (eds): Human Artificial Insemination and Semen Preservation. New York, Plenum Press, 1979, pp 197–210.

46   Camposfilho N, Franco ELF: Microcomputer-assisted multivariate survival data analysis using Cox's proportional hazards regression model. Comput Methods Programs Biomed 1990;31:223–228.

47   Colpi GM, Sagone P, Tognetti A, Campana A, Piffaretti-Yanez A, Balerna M: Linear and non-linear relationships between the 'swelling test' and conventional semen variables in men suspected of primary infertility. Human Reprod 1990;5: 600–605.

48   Avery S, Bolton VN, Mason BA: An evaluation of the hypo-osmotic sperm swelling test as a predictor of fertilizing capacity in vitro. Int J Androl 1990;13:93–99.

49   Rosenberg JL: Design of experiments for evaluating frozen sperm. J Androl 1990; 11:89–96.

50   Baker HWG: Management of male infertility; in Serio M, Waites G (eds): Recent Advances in Andrology. Serono Symposia Rev No 20. Rome, Ares-Serono Symposia, 1989, pp 81–92.

51   Comhaire F, de Kretser D, Farley TMM, Rowe PJ: Towards more objectivity in diagnosis and management of male infertility. Int J Androl, 1987, suppl. 7.

Dr. Marco Balerna, Andrology Laboratory, Gynecological Endocrinology Unit, 'La Carità Hospital', CH–6600 Locarno (Switzerland)

Colpi GM, Pozza D (eds): Diagnosing Male Infertility.
Prog Reprod Biol Med. Basel, Karger, 1992, vol 15, pp 19–26

The Usefulness of New Tests in the Evaluation of the Ejaculate

M. De Rosa, B. Amalfi, M. Mennitti, S. Zarrilli, G. Lombardi
Chair of Endocrinology, Second School of Medicine, University of Naples, Italy

The qualitative evaluation of the ejaculate has in recent years exploited new functional techniques. These methods have been developed to examine the ability of spermatozoa to reach a good ovum fertilization. This ability is not easy to study and it justifies the aphorism of some authors [1, 2]: 'pregnancy is the only irrefutable proof that a sperm can fertilize'. In spite of this, it is necessary to evaluate the male factor in case of lack of pregnancy after 12 months, at least, of free intercourses.

In this study, the features evaluated until the present time are strongly aided by new techniques. The sperm count, morphology and motility can show the fertilizing ability of the ejaculate. Among the sperm qualitative tests we chose (1) the eosin test (ET); (2) the hypo-osmotic swelling (HOS) test, and (3) the penetration test in bovine cervical mucus (CMPT) for their reliability and simplicity so that they can be performed as routine procedures.

The ET [3] is based upon the property of a supravital staining, the eosin Y, to penetrate in the unviable sperm cells that become bright yellow or red while the live sperm remain unstained. This method permits to calculate the percentage of live sperm without considering their motility.

The HOS test [4] shows the functional integrity of sperm membrane by its capacity to allow the water-active transport from the extracellular to the intracellular milieu, to balance the osmotic difference, when the spermatozoa are treated with hypo-osmotic solution. When the water enters the cytoplasm, some tail modifications, consisting in their twisting and/or swelling, occur.

The CMPT [5] shows the sperm proceeding ability in the bovine cervical mucus that has the same biophysical and biochemical properties of the human cervical mucus during the preovulatory period.

Materials and Methods

Seminal fluid samples were obtained by masturbation, after 3 days of abstinence in 61 subjects aged between 19 and 35 (mean 26 ± 6.2). The subjects were divided into 7 subsets: (1) normal controls who have induced a pregnancy in their partner (n = 12); (2) subjects with surgically treated cryptorchidism (n = 3); (3) grade 1–2 varicocele (n = 10); (4) grade 3–5 varicocele (n = 5); (5) asymptomatic bacteriospermia (n = 22); (6) idiopathic oligoasthenospermia (n = 4); (7) surgically treated varicocele (n = 5) (tables 1, 2).

Table 1. Sperm motility and qualitative tests in 12 normal subjects (mean ± SD)

	Normal controls
Sperm motility in 1st hour, %	64.16 ± 5.14
Forward progression, %	50.41 ± 8.64
HOS test, %	60.83 ± 6.33
ET, %	67.08 ± 6.89
CMPT, mm	32.33 ± 4.07

Table 2. Sperm motility and qualitative tests in 49 oligospermic patients with different gonadal pathologies (mean ± SD)

	Patients
Sperm motility in 1st hour, %	40.71 ± 17.19
Forward progression, %	23.80 ± 16.57
HOS test, %	36.90 ± 19.52
ET, %	53.33 ± 17.41
CMPT, mm	20.57 ± 10.90

The grades of varicocele were established after clinical examination and echo-Doppler fluximetry according to Maidoveir-Popesca. The asymptomatic bacteriospermia were confirmed after two semen cultures. The surgically treated patients had undergone surgery at least 18 months before. After collection, the ejaculates were allowed to liquify at room temperature. After the sperm count, motility was evaluated in a Makler counting chamber ($40 \times$), while the sperm morphology was evaluated after dilution (1:1) in phosphate-buffered saline and Giemsa. The ET was done according to Eliasson and Treichi [3] within 60 min after ejaculation with 0.1 ml of staining solution and 0.1 ml of ejaculate; after air desiccation the slide was observed at $40 \times$. The HOS test was performed with the solution prepared according to Jeyendran et al. [4], allowing 30 min at $37\,°C$. The CMPT was done with commercially available kits (Penetrak-Serono) and incubated for 60 min at $37\,°C$. All the patients showed an oligospermia with different grades of asthenospermia.

Statistical analysis was done by the correlation coefficient applied to the CMPT versus the other parameters.

Results

The statistical analysis results show a significant correlation ($p <$ 0.05) among the CMPT and the other parameters between the controls and the oligospermic subjects, except for the ET (fig. 1–4). The results of the various parameters in the different subsets of patients are reported in table 3.

Table 3. Sperm motility and qualitative tests in 49 oligospermic patients, taking in account the gonadal pathology (mean ± SD)

	Group[1]					
	A (n = 10)	B (n = 5)	C (n = 5)	D (n = 3)	E (n = 22)	F (n = 4)
Sperm count/mm³ ($\times$ 10)	46.1±13.4	11.0±7.5	47.0±19.8	18.5±8.3	37.5±15.3	18.5±7.5
Sperm motility in 1st hour, %	49.5±13.4	11.0±8.9	48.0±14.8	43.3±11.5	37.2±22.4	32.5±22.1
Forward progression, %	37.0±6.7	6.0±3.9	24.0±15.1	30.0±0	19.0±9.2	30.0±10.0
HOS test, %	42.0±14.9	29.0±13.0	38.0±25.8	15.0±5.0	36.0±10.1	16.2±16.0
ET, %	59.0±8.7	38.0±19.2	54.0±20.7	53.3±11.5	50.0±21.4	42.0±5.0
CMPT, %	22.3±8.4	17.2±9.4	27.2±5.3	23.5±6.7	18.5±10.9	9.6±8.7

[1] Group A: grade 1–2 varicocele; group B: grade 3–5 varicocele; group C: surgically treated varicocele; group D: surgically treated cryptorchidism; group E: asymptomatic bacteriospermia; group F: idiopathic oligo-asthenospermia.

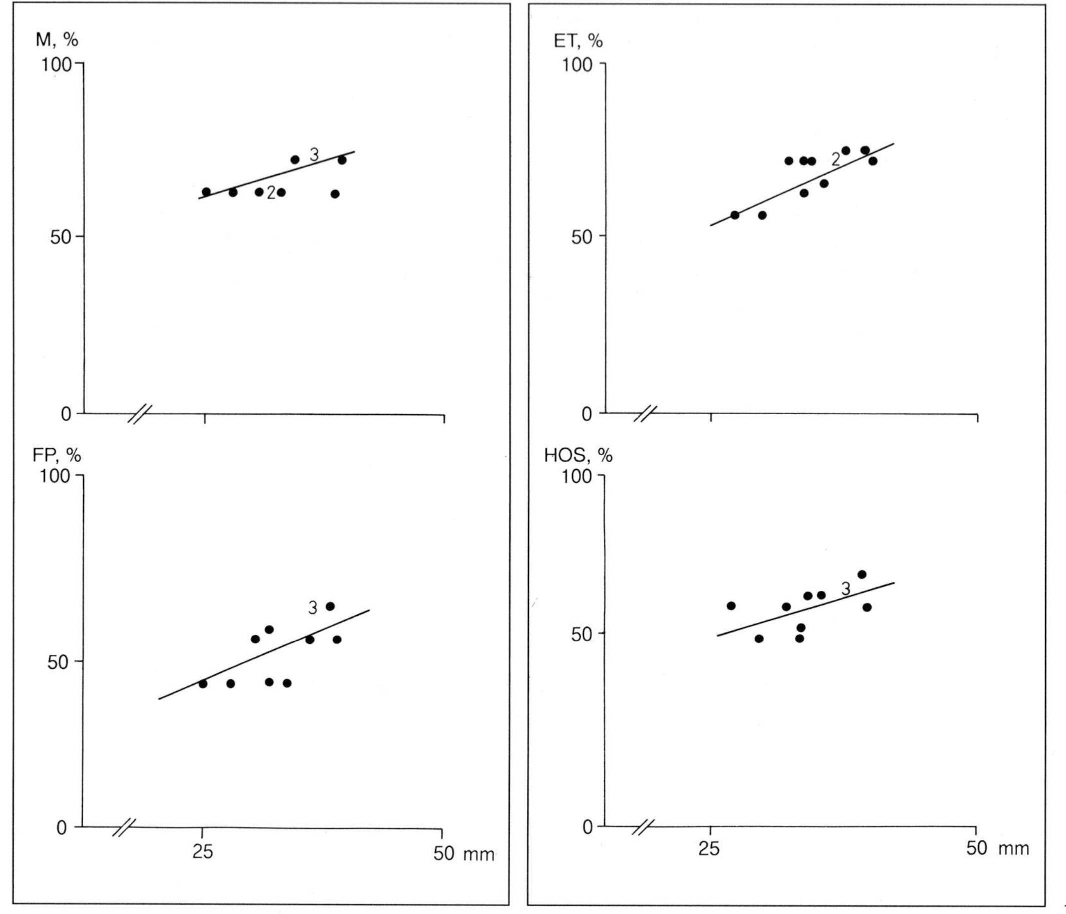

Fig. 1. Relationship between CMPT versus 1st hour motility (M) and versus forward progression (FP) in 12 normal subjects (as %, r = 0.62 and r = 0.60). The numbers indicate the cases that have the same value.

Fig. 2. Relationship between CMPT versus ET and HOS test in 12 normal subjects (as %, r = 0.81 and r = 0.56). The numbers indicate the cases that have the same.

Discussion

The swelling pictures are typical of normal spermatozoa after hypo-osmotic stress, normally over 60% [7], while over 50% are pathological. This phenomenon is present in 1–5% of normal ejaculates, though it is easy to see swollen tails according to different authors [8, 9]. It has recently

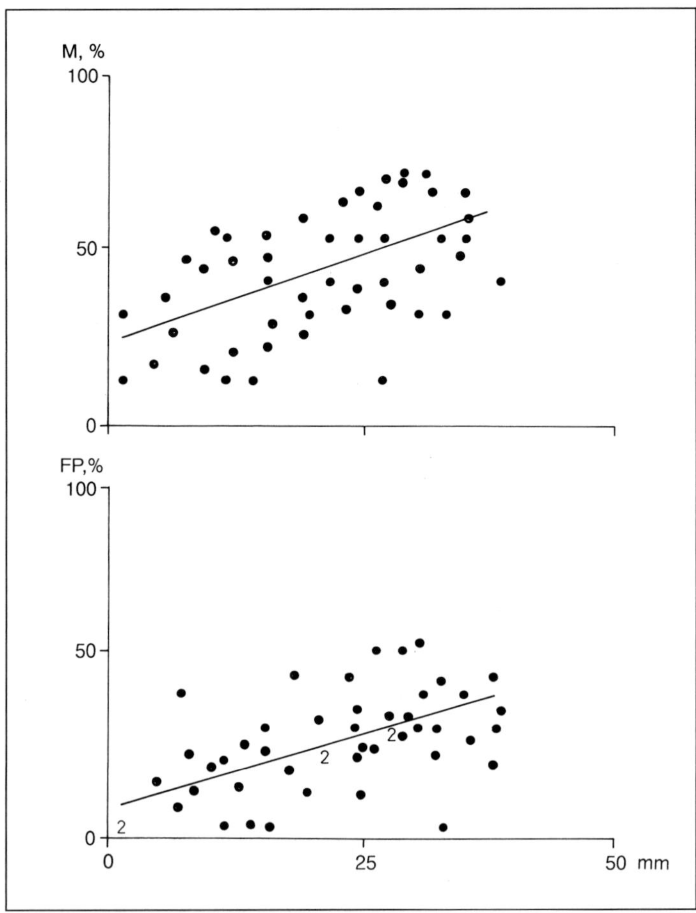

Fig. 3. Relationship between CMPT versus 1st hour motility (M) and versus forward progression (FP) in 49 patients affected with oligospermia (as %, r = 0.51 and r = 0.53). The numbers indicate the cases that have the same value.

been reported that the percentage is over the 5% outlined by Jeyendran et al. [4]. In our experience, based upon the sperm morphology observation or the examination coupling the HOS test and the ET, it has been demonstrated, according to Jeyendran et al. [4], that the swollen spermatozoa are live and then the swelling is a functional state.

The spontaneous swelling is present in subjects affected with genital infections, particularly for gram-positive cocci [11], *Micoplasma* or *Chla-*

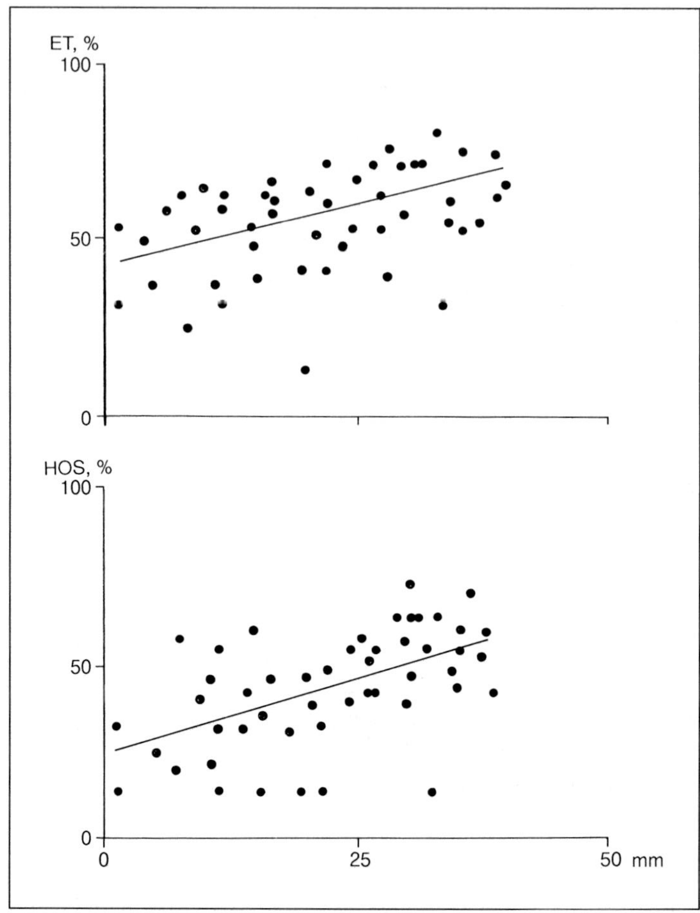

Fig. 4. Relationship between CMPT versus ET and HOS test in 49 patients affected with oligospermia (as %, r = 0.34 and r = 0.41).

mydia trachomatis [12]. The same infective damage reduces the capacity of swelling from osmotic stress, as pointed out by our data.

The CMPT outlines the seminal characteristics (motility, viability and morphology) [13] and, as the swelling test [14], showed a better correlation with the other parameters than the classical parameters [15]. Nevertheless, we must take into account the role seminal fluid plays in cervical mucus penetration, whether positively [16] or negatively when it is modified for genital tract inflammation [17]. Regarding the results of bacteriospermia and according to others [17–19], it could be outlined that the seminal fluid is so damaged to be responsible for the deficient sperm penetration in the cervical mucus.

In conclusion, the usefulness and the reliability of these tests seem to be confirmed by our experience. Furthermore, they give useful information about the damage of the sperm cells in different pathologies. This information could be useful to begin the treatment and, probably, to choose eventually the artificial insemination.

References

1   Alexander NJ: A new test of sperm function: in vitro CMP. Contemp Obstet Gynecol 1983;Feb:1–8.
2   Blasco L: Clinical test of sperm-fertilizing ability. Fertil Steril 1984;41:177–192.
3   Eliasson R, Treichi L: Supravital staining of human spermatozoa. Fertil Steril 1971; 22:134.
4   Jeyendran RS, van der Ven HH, Perez-Pelaez M, Crabo BG, Zaneveld LJD: Development of an assay to assess the functional integrity of the human sperm membrane and its relationship to other semen characteristics. J Reprod Fertil 1984;70:219–228.
5   Moghissi KS, Segal S, Meinhold D, Aronow SJ: In vitro sperm cervical mucus penetration: Studies in human and bovine cervical mucus. Fertil Steril 1982;37:823–827.
6   Foresta C, Indino M, Scanelli G, Mioni R, Zorzi M, Trevisan L, Scandellari C: Considerazioni sul significato clinico di nuovi parametri funzionali degli spermatozoi in soggetti fertili ed infertili; in Foresta C, Scandellari C (eds): Inseminazione Artificiale Omologa. Padova, Libreria Cortina, 1987, pp 85–117.
7   Check JH, Nowroozi K, Wu CH, Bollendorf A: Correlation of semen analysis and hypo-osmotic swelling test with subsequent pregnancies. Arch Androl 1988;20:257–260.
8   Menchini-Fabris GF, Marcellino I, Cavirano G, Valiani S, Lombardi M, Giannotti P, Di Coscio M, Cicalese V, Olivieri L: Morfologia spermatozoaria nell'eiaculato normale e patologico (lo spermiocitogramma nell'uomo). Fisiopat Rip 1983;1:169–173.
9   Schwartz D, Mayaux M-J, Spira A, Moscato ML, Jouannet P, Czyglik F, David G: Semen characteristics as a function of age in 833 fertile men. Fertil Steril 1983;39:530–535.
10 Saltarelli O, Chelo E, Livi C, Beretta G: A new criterion for the more exact evaluation of the swelling test. Acta Eur Fertil 1989;20:87–97.
11 Baccetti B, Burrini AG, Magnano AR, Piomboni P, Renieri T, Sensini C, Bilenchi R, Rossolini G, Setacci C, Zanchi S: Bacteria and spermatozoa; in Serio M (ed): 4th International Congress of Andrology, 1989, abstr 333.

12 Serio M, Paliaga L, Bonfanti L, Casano R, Natali A, Menchi I, Gianti E: Le prostatiti. Aggiornamento Med 1990;14:1.
13 Bergman A, Amit A, Menachem PD, Homonnai ZT, Paz GF: Penetration of human ejaculated spermatozoa into human and bovine cervical mucus. Correlation between penetration values. Fertil Steril 1981;36:363–367.
14 Check JH, Winkel CA, Shanis BS, Bollendorf A, Epstein RH: The hypo-osmotic swelling test as a useful adjunct to the semen analysis, to predict fertility potential; in Serio M (ed): 4th International Congress of Andrology, 1989, abstr 128.
15 Alexander NJ: Evaluation of male infertility with an in vitro cervical mucus penetration test. Fertil Steril 1981;36:201–208.
16 Overstreet JW, Coats C, Katz DF, Hanson FW: The importance of seminal plasma in sperm penetration of human cervical mucus. Fertil Steril 1980;34:569–572.
17 Eggert-Kruse W, Gerhard I, Hofmann H, Runnebaum B, Petzoldt D: Influence of microbial colonization on sperm–mucus interaction in vivo and in vitro. Hum Reprod 1987;2:301.
18 Nikkanen V, Gronnos M, Suominen J, Multamaki S: Silent infertility in male accessory genital organs and male infertility. Andrologia 1979;11:236.
19 Panidis D, Rousso D, Matalliotakis J, Manikas J, Kalogeropoulos A: Coexistence of spermatozoa morphological abnormalities in the semen of men with infection of the accessory genital glands; in Serio M (ed): 4th International Congress of Andrology, 1989, abstr 59.

Dr. Michele De Rosa, via Belsito 19, I–80123 Napoli (Italy)

Colpi GM, Pozza D (eds): Diagnosing Male Infertility.
Prog Reprod Biol Med. Basel, Karger, 1992, vol 15, pp 27–48

Recent Developments in Histophysiology of Male Accessory Sex Glands

Gerhard Aumüller, Ulrich Rausch, Jürgen Seitz
Department of Anatomy and Cell Biology, Philipps-Universität, Marburg, FRG

The modification of the sperm cell surface during the transit of the cells through the epididymis is well established [Bedford, 1963], and a number of specific proteins with discrete functions has been identified [e.g. by Acott and Hoskins, 1981; Rifkin and Olson, 1985]. No such detailed knowledge exists about the action on spermatozoa of the several proteins secreted by the accessory sex glands, i.e. prostate and seminal vesicles. The secretions from male accessory sex glands interact with each other and with spermatozoa, their predominant functions being semen gelation, coagulation and liquefaction, coating and decoating of spermatozoa, ionic and metabolic exchange reactions between seminal plasma and spermatozoa, and interaction with cervical mucus [Mann and Lutwak-Mann, 1981]. While the method of split-ejaculates yielded only rough estimates of the nature of various compounds and their respective sources, two-dimensional gel electrophoresis (2D-PAGE) of semen or isolated glandular secretion [Lilja and Abrahamsson, 1988] or of organ homogenates [Lee and Sensibar, 1987] has led to the identification and molecular and functional characterization of several proteins, both in humans and experimental animals [Gerhardt et al., 1983]. Immunosuppressive material [James and Hargreave, 1984], sperm-motility-blocking agents [de Lamirande and Gagnon, 1983], antifertility compounds and a sperm-binding protein [Abrescia et al., 1985] have been recently described in human semen. The search in forensic medicine for specific markers of human semen, on victims of sexual assault, has revealed a variety of different proteins, one of which turned out to be the essential gel-forming substance delivered from seminal vesicles into semen [semenogelin: Lilja and Laurell, 1984; Herr et al., 1986, 1989; McGee and Herr, 1987]. The differences in clotting and liquefaction observed in human semen can now be determined quantitatively, since both the clotting and liquefying system (prostate-specific antigen, PSA) [Wang et al., 1979] have now been isolated and characterized. Another important clinical aspect of male accessory sex glands is the pathological enlargement of the prostate (benign prostatic hyperplasia, BPH) observed in more than 80% of the male population before the age of 80 years, and far less common than BPH but far more deadly, cancer of the prostate, the third most frequent cause of male tumor deaths.

Although quite a number of secretory proteins are known in the human prostate, only three [Lilja and Abrahamsson, 1988] have gained major importance as diagnostic tools, namely prostatic acid phosphatase (PAP), PSA and β-microseminoprotein (P-MSP). The nature of the latter is controversial [Dubé et al., 1987] since it shows partial sequence homology with β-inhibin [Garde and Sheth, 1989]. In view of pathogenic mechanisms in the human prostate, additional secretory proteins such as prostatic growth factors [Hierowski et al., 1987; Story et al., 1987] and hormones [relaxin: Weiss, 1989] have aroused the interest of both biochemists and pathologists, since they were thought to be of particular significance in the regulation of prostatic growth [Tenniswood, 1986]. One essential question is the functional role of distinct proteins such as caltrin [Comte et al., 1986] or semenogelin (SG) at the subcellular level in spermatozoa. The relationship between different proteins from the rat prostate and seminal vesicles was carefully examined at either the DNA or the protein level by Kandala et al. [1985a, b], Williams et al. [1985], Fawell and Higgins [1987] and Fawell et al. [1986]. While there was only a limited relationship between different species, upstream homologies in the genes for SVS IV and the C3 component of PBP were strikingly comparable in a 30 nucleotide sequence that deserves further consideration as a potential site involved in the androgen-regulated expression of these genes [Kandala et al., 1985a, b]. Perhaps several questions on secretory proteins from the accessory sex glands will be answered from studies at the DNA level, where the molecular structure and the functional activity of the respective protein failed to give a clue on its true functional role, for example, prostatic acid phosphatase.

Process of Semen Coagulation and Liquefaction

Human seminal fluid is known to contain a great variety of secretory proteins which exert different biochemical functions such as semen coagulation and liquefaction, regulation of calcium fluxes and immunological mechanisms (table 1). Following ejaculation in the human, a soft, jelly-like coagulum of porous and lamellated ultrastructure is formed within 5 min and then dissolves within a period of 5–20 min [Tauber et al. 1980]. Semen coagulation has been considered to result from the activities of prostatic clotting enzymes on proteins derived from the seminal vesicles [Mann, 1964] and liquefaction from the action of fibrinolytic, proteolytic and collagenase-like peptidase activities. Three different proteinases are well known in human seminal plasma: pepsinogen, plasminogen activator, and a neutral proteinase called seminin [Tauber et al., 1976]. Acidic proteinases have been found in human semen by Lundquist [1952] along with an acid proteinase inhibitor [Minakata and Asano, 1985]. Recent studies have identified human prostatic gastricsinogen rather than a pepsin-like enzyme as a precursor of seminal fluid-derived acid proteinases [Reid et al., 1984]. Both tissue-type and urokinase-type plasminogen activator are present in the human prostate [Kirchheimer et al., 1984]. Plasminogen activator has

Table 1. Interaction between the male accessory secretions and spermatozoa [Mann and Lutwak-Mann, 1981; Aumüller and Seitz, 1990a, b]

Copulatory plug formation (rodents)
Coagulation and liquefaction (human)
Gelation of semen (boar)
Coating and decoating of spermatozoa
Sperm-coating antigens
    Blood serum proteins
    Antigens common to semen and other body fluids
    Seminal plasma-specific proteins
Ion exchange reactions
    Calcium flux regulation
    Motility inhibitors
    Motility inductors
External functions ('transsecretion')
    Interaction with uterine structures
    Immunomodulation
    Embryonic development (?)

---

been related to degradation of fibrin-like proteins, formed during initial proteolytic process [Tauber et al., 1976]. More recently, Tauber et al. [1980] have reported that the addition of partially purified seminin is able to enhance liquefaction of the seminal plasma coagulum. Koren and Lukac [1979] have studied the effects of ions on semen liquefaction and found that copper inhibits and EDTA enhances liquefaction. Based on these observations of ionic interactions in clot formation and liquefaction together with the rheological properties of the coagulum, Daunter et al. [1980] have suggested liquefaction of the glycoprotein-metal ion complex of the coagulum to result from reduction of the metal ions by L-ascorbic acid. Polak and Daunter [1989] have discussed that removal of hydrogen peroxide generated by the oxidation of L-ascorbic acid requires the activity of glutathione peroxidase and glutathione reductase which have been identified in the human seminal plasma. The significance of seminal vesicle proteins and PSA, a prostate-derived kallikrein-type serine proteinase in semen coagulation and liquefaction has recently been fully elucidated by Lilja et al. [1989], and has been confirmed by Herr et al. [1989]. Lilja and Laurell [1984, 1985] found the predominant protein in the seminal vesicle to constitute the structural protein of the seminal coagulum (designated a HMW-SV-protein). Using a monoclonal antibody against the MHS-5 antigen from seminal fluid, Herr et al. [1986] reported the presence of the antigen in semen obtained from postvasectomy patients. Evans and Herr [1986] were able to localize the antigen in human seminal vesicle epithelium. Using an electron microscopic approach, Herr et al. [1989] have found a positive immunoreaction with MHS-5 antibody in the electron-dense core of the secretory granules of seminal vesicle glandular cells. It shows an immunolocalization analogous to SVS II in rat seminal vesicles [Aumüller and Seitz, 1986] which is the major constituent of the copulatory plug in rats. Obviously, the protein and its fragments observed in semen described by Herr et al. [1989] is basically identical with that described by Lilja et al. [1989]. Studying the distribution of the antigen in epididymal and ejaculatory sperm, Evans and Herr [1986] concluded that the MHS-5 antigen may be considered a 'sperm-coating antigen' like other proteins described previously [Wichmann et al., 1989].

Lilja [1985] was the first to describe the cleavage of the predominant seminal vesicle protein by PSA. PSA is a proteolytic enzyme [Ban et al., 1984]. It is a 33-kD glycoprotein and is present at high concentrations in human prostatic fluid. This protein and the closely related canine arginine esterase [Dubé et al., 1985] are kallikrein-like proteases. The human PSA has been shown to cleave a kininogen-related substrate in seminal vesicle secretion. SG is dissociated into three subunits of some 52, 71, and 76 kD after reduction. There are few discrepancies concerning the size of the fragments formed during liquefaction [Lilja, 1985; McGee and Herr, 1987]. The predominant product formed during PSA cleavage is a 5.8-kD basic protein whose amino-terminal portion is similar to the histidine-rich region of bovine high molecular weight kininogen [Lilja and Jeppsson, 1985].

SG, fibronectin, and lactoferrin are the predominant secretory products of the human seminal vesicles; the total concentration of SG in the secretion being about ten times that of fibronectin and lactoferrin. Lilja et al. [1987] provided evidence that during the immediate postejaculatory phase fibronectin and SG are the predominant structural proteins of the seminal gel, whereas lactoferrin remains soluble. According to these authors, there are no disulfide or transglutaminase-linked covalent bonds in the human seminal coagulum [for discrepant findings, see Koren and Lukac, 1979]. Lilja et al. [1987] suggest that components of the prostatic secretion induce the polymerizing reaction between fibronectin and SG in a manner similar to that of the fibronectin interaction with platelets and fibrin during blood clotting. The gradual release of immunoreactive fibronectin into the seminal fluid observed by Lilja et al. [1987] during the liquefaction phase is apparently the result of progressive cleavage of gel-bound fibronectin effected by PSA. Among the fragments formed during the late phase of liquefaction, an abundant basic SG fragment has been isolated by Lilja and Jeppsson [1985]. This basic fragment is composed of 52 amino acid residues and shares structural elements with α-inhibin-92 and α-inhibin-31, two additional proteins isolated from human liquefied semen [Seidah et al., 1984; Lilja and Jeppsson, 1985]. Lilja et al. [1989] therefore concluded that both α-inhibin-92 and α-inhibin-31 may derive from SG during fragmentation. The physiological importance of the inhibin-like activity of the two SG fragments remains to be established. The same applies to β-inhibin (β-MSP) and its possible relationship to gastricsin [Lija and Abrahamsson, 1988]. The interplay between PSA and SG may be significant in the regulation of semen viscosity. A number of fertility disorders have been related to the fact that the seminal coagulum forming during ejaculation fails to liquefy and the ejaculate remains viscous. Dubé et al. [1989] have studied the concentration of immunoreactive PSA in seminal plasma from patients with different semen volumes and viscosities. They found no reduced immunoreactive PSA in seminal plasma with high viscosity. This may indicate that either PSA immunoreactivity is not equivalent with PSA enzyme activity (which may be regulated by seminal inhibitors) or that different PSA preparations yield rather different chemical products. The PSA preparation of Dubé et al. [1989], for instance, was unable to hydrolyze arginine-containing synthetic substrates, while the preparation of Lilja [1985] hydrolyzed preferentially arginine- and lysine-containing substrates. Finally, one must not forget the possibility that although PSA is the essential protease for SG hydrolysis, other enzymes may contribute to the liquefaction process.

An interesting question is the species specificity of human seminal proteins. We have used polyclonal antibodies against SG isolated from human seminal vesicle secretion and PAP, β-MSP and PSA derived from human prostatic fluid, as well as a monoclonal antibody against β-MSP for immunocytochemical detection of the respective antigens in different organs from different species. SG immunoreactivity was detected in the epithelium of the pubertal and adult human and in monkey seminal vesicle, ampulla of the vas deferens and ejaculatory duct. PAP, β-MSP and PSA immunoreactivities were detected in the pubertal and adult human prostate and the cranial and caudal monkey prostate. With the exception of a weak PSA immunoreactivity in the proximal portions of the ejaculatory duct, none of the latter antisera reacted with seminal vesicle, ampullary and ejaculatory duct epithelium. Among the nonprimate species studied (dog, bull, rat, guina pig) only the canine prostatic epithelium displayed a definite immunoreactivity with the PAP antibody and a moderate reaction with the PSA antibody. No immunoreaction was seen in bull and rat seminal vesicle and canine ampulla of the vas deferens with the SG antikody. The same was true for the (ventral) prostate of rat, bull and dog for β-MSP. The epithelium of the rat dorsal prostate showed a slight cross-reactivity with the monoclonal antibody against β-MSP and one polyclonal antibody against PSA. The findings indicate a rather strict species-dependent expression of human seminal proteins which show some similarities in primates, but an only marginal relationship to species with different physiology of seminal fluid.

Sperm-Binding Proteins

In addition to sperm-coating antigens, mostly derived from epididymis, an increasing number of sperm-binding proteins has been detected in seminal plasma. These proteins are related to different steps of sperm capacitation, apparently by their interaction with ions such as calcium and with each other.

The human prostate produces calcitonin [Sjöberg et al., 1980] that binds to the middle piece and the neck region of spermatozoa [Forestà et al., 1986], but its significance in the regulation of calcium fluxes in the middle piece region of spermatozoa is not clear. Calcium transport in sperm plays a significant role in the acrosome reaction [Meizel, 1985] and perhaps in motility [Tash and Menas, 1982]. During their transit through the epididymis, spermatozoa develop considerable changes in the kinetics of calcium transport. The ejaculated sperm are relatively impermeable to calcium [Babcock et al., 1979; Rufo et al., 1982], but permeability increases during their passage in the female reproductive tract [Meizel, 1985]. Various components were suggested to be involved in the transport of calcium into the spermatozoa; these include a Ca^{2+}/Mg^{2+}-ATPase, a Na^+/K^+-ATPase, a Na^+/K^+-antiporter and calmodulin (CaM), all of which have been demonstrated in isolated sperm [Sidhu and Guraya, 1989]. As to the functional significance of these enzyme systems, several divergent views have been reported. Sidhu and Guraya [1989] for example propose that in buffalo sperm, a CaM-like protein in seminal plasma binds to the sperm during ejaculation and stimulates Ca^{2+} extrusion by activating a Ca^{2+} pump. Tash et al. [1988] demonstrated that the inhibition of reactivated sperm motility by calcium was correlated with inhibited protein phosphorylation. The inhibition of phosphorylation (of 14 phosphoprotein substrates such as axokinin) was found to be catalyzed by a CaM-dependent phosphatase that was effective to modify sperm motility in a manner similar to that produced by Ca^{2+} without added phosphatase. As far as we know, no studies have as yet been performed using purified PAP on epididymal sperm to study a possible dephosphorylating activity on sperm phosphoproteins.

A protein antagonizing the effects of CaM in bovine semen has been studied by Gietzen and Galla [1985] and Comte et al. [1986]. It was found to be identical with the previously described seminal plasmin isolated from bull seminal plasma. This is a low molecular protein basic protein with a pI value of 9.8 [Reddy and Bhargava, 1979]. Seminal plasmin consists of 48 amino acids, the molecular weight being 6,385 kD. Amino acid analysis showed that it is rich in lysine and arginine [Theil and Scheit, 1983]. Owing to a random distribution of hydrophobic amino acids, seminal plasmin is highly water-soluble and shows no tendency of self-aggregation in an aqueous medium. This protein binds strongly to RNA polymerase and reverse transcriptase [Scheit et al. 1979] and inhibits RNA synthesis in bacteria and fungi [Scheit et al., 1985]. Immunohistochemically, it has been localized within the epithelium of bovine seminal vesicles [Shivaji et al., 1984; Aumüller and Scheit, 1987].

Gietzen and Galla [1985] recently reported that seminal plasmin (like other water-soluble, monomeric proteins displaying the characteristics mentioned) antagonizes specifically and with high potency the function of CaM, the major Ca^{2+} vector in eukaryotic cells. Comte et al. [1986] established that seminal plasmin forms a complex with calmodulin in the ratio of 1:1 and this complex is Ca^{2+}-dependent and urea-resistant. These papers are consistent with the findings of Lewis et al. [1985] and Sitaram et al. [1986] that seminal plasmin and caltrin are the same protein. Caltrin has initially been described by Babcock et al. [1979] and characterized by Rufo et al. [1982] and Lewis et al. [1985]. San Agustin et al. [1987] have performed immunofluorescence studies showing that caltrin binds to the plasma membrane over the acrosome and principal tail regions of bovine spermatozoa, but not to the postacrosomal area or the midpiece. In in vitro experiments, calcium influx into epididymal spermatozoa was found to be prevented by freshly prepared caltrin from bovine seminal plasma. Contrary to that, enhanced calcium uptake into these cells was measured if older preparations of seminal plasmin were used. San Agustin et al. [1987] postulated that during the early part of sperm transit through the female reproductive tract, caltrin bound to the sperm plasma membrane protects the sperm cells from calcium influx. It could be hypothesized that factors present in the oviduct, where spermatozoa meet the eggs, may cause caltrin to change from an inhibitor to an enhancer of calcium uptake. Capacitation, i.e. calcium-influx initiated acrosomal reaction and hyperactive motility then could take place.

The existence of caltrin-like proteins in the reproductive tract of the guinea pig has been reported by Coronel et al. [1988]. In contrast to the basic character of bovine caltrin, guinea pig's caltrin-like protein is acidic in nature and has three isoforms with pI values of pH 5.6, 6.0 and 6.2 respectively. These data and the reported thermostability of the guinea pig polypeptide indicate differences in amino acid composition between the proteins from these two species.

The findings of San Agustin et al. [1987] relate bovine sperm motility closely to the functions of caltrin. In the bovine genital tract, however, seminal plasma is only the secondmost active fluid (the first being epididymal fluid) in producing forward motility by increasing the number of moving epididymal spermatozoa [Brandt and Hoskins, 1980]. Hoskins et al. [1979] have performed a series of elegant studies on motility induction of spermatozoa resulting in the identification of an epididymal protein that acts as a forward motility protein (FMP). Acott and Hoskins [1981] demonstrated that FMP is present both in bovine epididymal and seminal fluid. Agrawal and Vanha-Perttula [1987] described 'vesiculosomes' released from bovine seminal vesicle epithelium that induce hyperactivation of sperm motility and the acrosomal reaction in epididymal spermatozoa when suspended in Ringer's medium containing those particles. In bovine semen, a low molecular weight (13,000) neutral protein is present, that constitutes the majority of secretion [Esch et al., 1983; Scheit, 1986].

As has been demonstrated by cell-free translation experiments [Kemme et al., 1986], this neutral protein is synthesized as a 18-kD precursor which is processed to the definitive 13-kD form and secreted during ejaculation. Recently, we have demonstrated immunohistochemically the absence of this protein in the epithelium of seminal vesicle of calf [Aumüller et al., 1988]. After androgen stimulation following puberty, it appears in the epithelium of seminal vesicle and ampulla. No immunoreactivity is found elsewhere in the bovine male genital system. Intense immunoreactions were displayed by ejaculated spermatozoa as well as spermatozoa present in the ampulla of vas deferens and the urethra. Immunoreactions were confined to the neck region and middle piece of the sperm, while the principal piece of the tail as well as the sperm head were nonreactive. Using the immunogold technique (pre-embedding staining after removal of the plasma membrane) at the ultrastructural level, major protein was visualized close to the mitochondria of the midpiece and within the implantation fossa of the tail at the neck. Epididymal sperm extracts separated by SDS-PAGE and transblotted onto nitrocellulose sheets demonstrated, after incubation with gold-labeled major protein, a protein duplet with a molecular weight of 65 and 67 kD, respectively. This pattern of staining appears to represent the binder of major protein underneath the sperm surface. Binding of major protein to this acceptor site is regarded as a physiological event that may be related with the onset of hyperactivated sperm motility. Our in vitro experiments indicate a pronounced structural lability of the plasma membrane in the middle piece region of the bovine spermatozoa. This was particularly obvious when calcium-free solutions were used. We therefore conclude that different compartments or domains of the plasma membrane of bovine spermatozoa are differentially sensitive to changes in calcium concentrations and these differences can be regulated by seminal proteins such as caltrin and major protein.

Increment and inhibition of sperm motility by seminal compounds are closely related. De Lamirande and Gagnon [1983] have isolated a seminal plasma motility inhibitor (SPMI) which decreases the percentage of motile spermatozoa in a dose- and time-dependent manner. According to these authors, the data suggest that SPMI could play a significant role in cases of infertility caused by asthenospermia.

Secretory Proteins from Rat Male Accessory Sex Glands: Biochemical and Functional Characterization

During the last 5 years we have focused our studies on those factors in semen which actively or passively participate in semen coagulation and liquefaction. It is known from the work of Williams-Ashman and co-workers [Williams-Ashman, 1984] that in guinea pig mainly the seminal vesicles (SV) and the anterior prostate (= coagulating gland, CG) produce the essential factors: the SV provides monomeric proteins and the CG the linking enzyme (transglutaminase).

To answer the question whether in rat semen an identical mechanism is operating as it is in guinea pigs, we had to isolate the respective proteins in highly purified and native form (denatured proteins have been used for antibody production by Fawell et al. [1986]. The isolated proteins were then further characterized with respect to their molecular properties and their functional role. The specific objective of our study was to reconstitute semen coagulation in vitro and to compare with the authentic mechanism in vivo (fig. 1). We hoped to get insight into the completely obscure mechanism underlying coagulation of human semen.

Secretory Proteins from Rat SVs

Definition and distribution: Using SDS-PAGE of rat SV fluid, Ostrowski et al. [1979a] have identified 5 proteins which were numbered according to their respective electrophoretic mobility: SVS I (80 kD), SVS II (49 kD), SVS III (33 kD), SVS IV (16 kD) and SVS V (15 kD). Somewhat later, Ostrowski et al. [1979b] found the FAD-dependent enzyme sulfhydryloxidase (SOx, 65 kD) in seminal vesicle secretion.

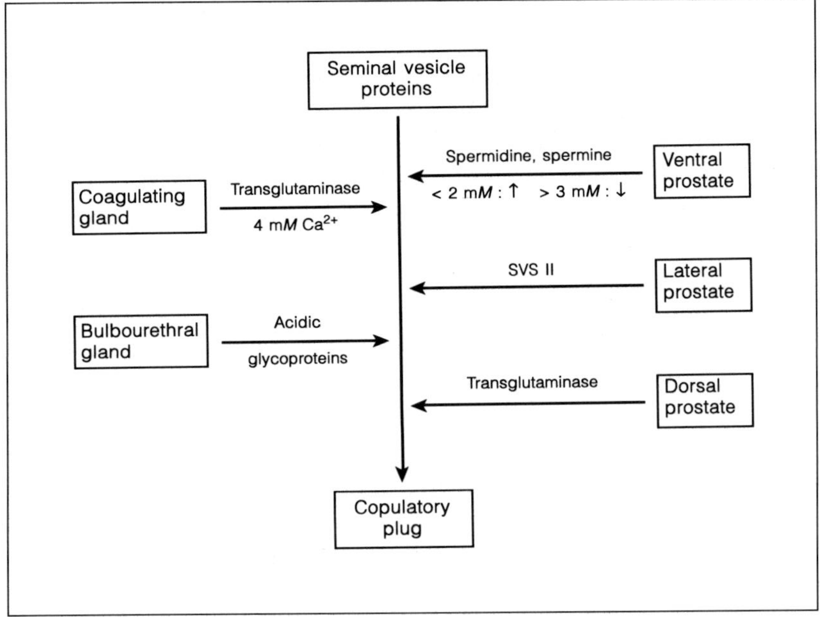

Fig. 1. Secretions from rat accessory sex glands contributing to the formation of the copulatory plug.

By means of high-resolution SDS-PAGE (gradient range 10–25% acrylamide) we were able to subfractionate SVS V into 3 proteins increasing the number of polypeptides up to 8. SOx was not generally present in SV secretion. During our studies on the influence of sexual activity on morphology and biosynthetic activity of the accessory sex glands [Aumüller et al., 1985], a heterogeneous protein pattern of SV secretion was observed; SOx was preferentially concentrated in SV secretion from sexually active animals. The enzyme is able to link monomeric SVS II and SVS IV with consumption of oxygen and forming nearly insoluble oligomers. Wagner and Kistler [1987] have additionally demonstrated the formation of heterogeneous adducts with SVS I and SVS III.

The enzymatically induced precipitation of SVS II obviously is the major reason for the nearly complete lack of this factor in the secretion of sexually active animals. The SV secretion of sexually inactive animals, where SOx activity is low, has an abundant proportion of SVS II in dimeric or trimeric form.

SVS II

SVS II solubility is strongly dependent on pH: at neutral to basic pH values the protein precipitates nearly completely. In the presence of 5 mM spermine and spermidine it can be redissolved by about 50% with glycine HCl buffer (pH 2.0) or 1 M aqueous solution of urea by more than 90%. The molecular weight of SVS II is 49 kD under denaturing and reducing conditions (SDS-PAGE). It is a dimer after native (pH 4.7) PAGE. After gel filtration on a G-150 column (in the presence of spermine and spermidine) it forms trimers, and the same is true after FPLC gel filtration (Superose 12) in the presence of acidic glycine buffer (pH 2.0), where a small proportion of hexamers is found as well. Isolated oligomers can be reduced to monomers by extensive dialysis against 10^{-2} M DTE, but these do not contain any biological activity. The isoelectric point of purified SVS II is about pH 10.5. The protein is the only one of SVS polypeptides showing a high affinity for lectins (either gold- or peroxidase-labeled) from *Tetragonolobus purpurea* (specific carbohydrate ligand: fucose) concanavalin A (glucose, mannose), *Ricinus communis* type I (galactosyl-N-acetylglucosamine) and type II (galactose) and *Triticum vulgaris* (glucose-N-acetylglucosamine, neu Nac). Specificity is demonstrated by inhibition of binding in the presence of unlabeled lectins as well as saturation with excess of specific sugars. This indicates SVS II as a glycoprotein. Gas chromatography of carbohydrate moieties has confirmed that finding. Amino acid analysis (kindly performed by Prof. Vanderkerckhove, Ghent) showed an atypically high content in Ser, Gly, Glu and Lys residues. The former basic residues make SVS II a nearly ideal substrate for transglutaminase (see below). The protein can be degraded at the numerous Lys residues by serine-type proteases (important during liquefaction).

SVS I

SVS I is present at higher amounts only in SV secretion of sexually inactive animals. It shows a tendency of precipitation and is cross-linked enzymatically by sulfhydryloxidase. Perhaps it is also incorporated into the copulatory plug. SVS II shows cross-reactivity with antibodies against SVS I, and vice versa. Peptide mapping with V$_8$-tryptic fragments separated by SDS gel electrophoresis demonstrated several comigrating peptides. This indicates a close molecular relationship between SVS I and SVS II, the larger polypeptide obviously being a dimer of SVS II.

SVS III–VII

No details are known at present of these polypeptides that are not glycosylated as well and that are isolated only exceptionally in native and homogeneous form. For instance, the functional role of SVS IV, the most concentrated protein in SV secretion, is completely unknown. As yet, SVS IV could be demonstrated by means of immunoelectron microscopy within the halo surrounding the secretory vacuoles of SV glandular cells [Aumüller and Seitz 1986]. SVS VI has a strong affinity for sulfhydryloxidase from which it can be separated only with major difficulties. There is, however, no molecular or immunological relationship between these two proteins.

Sulfhydryloxidase

This enzyme conjugates proteins by oxidation of SH groups in Cys residues forming disulfide bridges. Like several other oxidases it is dependent on FAD as cofactor [Ostrowski and Kistler, 1980]. Under denaturing conditions it has a molecular weight of 65 kD. We have developed a modification of the purification procedure described by Ostrowski and Kistler [1980]: after ammonium sulfate precipitation at 30–65% saturation of a tissue extract and desalting on Sephadex G-25 column, FPLC ion-exchange chromatography was performed on a Mono S column. SVS VI was subsequently removed by FPLC. Ion-exchange chromatography was performed on a Mono S column. SVS VI was subsequently removed by FPLC chromatofocusing with Mono P, elution of sulfhydryloxidase (SOx) occurred at a range of pH 8.6–8.3 close to the pI of the molecule, SOx activity was determined quantitatively with the aid of an oxygen electrode using DTE as a substrate. Alternatively, the molar ratio of FAD binding (1:1) to the molecule was used for quantification (direct monitoring in spectrophotometer at 450 nm). Both methods resulted in more than 95% purity of the protein which was confirmed by SDS-PAGE. The purified protein was injected into rabbits for antibody production. Immunoelectron microscopy using ultrathin cryosections demonstrated a labeling of the halo of the secretory granules in SV cells and in addition a specific labeling of the matrix of mitochondria [Seitz et al. 1987a]. The labeling of mitochondria was also observed immunohistochemically in kidney (cells of the distal tubule), salivary gland (striated duct cells), liver (periportal hepatocytes) and ovary (follicular epithelium, rete cells). Immunoprint analysis of extracts from the organs mentioned as well as from isolated liver mitochondria resulted in all cases in a cross-reactivity with a 65-kD protein indicating the close relationship between the secretory and the mitochondrial sulfhydryloxidase. Sulfhydryloxidase has been purified from the SV secretion of sexually active rats and is well characterized [Ostrowski et al., 1979b; Seitz et al., 1988]. The functional significance of this enzyme is not yet known, though its role in (1) regulating disulfide bonding on spermatozoa surface during epididymal maturation and storage, and (2) protecting spermatozoa against deleterious effects of low molecular weight thiols has been suggested [Chang and Zirkin, 1978]. Ostrowski et al. [1979a] suggested that thiol oxidase may serve to generate disulfide bonds in one or more of the major SV secretory proteins, but they have no evidence for such an assumption. Wagner and Kistler [1987] reported the ability of SOx from SVs to link monomeric SVS I, II and III to form homo- and hetero-oligomers in the presence of oxygen.

Using polyclonal antibodies raised against a purified preparation of SOx, we observed that SOx is specifically associated with the germ cells in the testis of rat and hamster and that it appears in a stage-dependent manner. We observed that germ cells from pachytene spermatocytes stage onwards have this enzyme. Initially, the SOx immunoreactivity is low in pachytene spermatocytes at stage I, but slowly increases and is found to be maximum in pachytene spermatocytes at stages VIII–IX in rat and at stages III–IV in hamster and remains the same up to diplotene/diakinesis stage. Thereafter, it seems to be less in spermatids though uniformly distributed in the cells immediately below the plasma membrane. The appearance of SOx in germ cells seems to be directly related to the appearance of 'condensed' mitochondria in germ cells [de Martino et al., 1979]. The earliest sign of formation of 'condensed' mitochondria is observed in early spermatocytes, but only few mitochondria are involved in this process. In zygotene and early pachytene spermatocytes, this biological process is seen in numerous mitochondria. The observed (1) increase in SOx in meiotic cells, (2) cytoplasmic distribution pattern in spermatocytes and spermatids (steps 1–8), (3) aggregation of SOx immunoreactivity into a small ball-like structure in maturating spermatids (steps 15–18) and (4) retention of SOx in the residual bodies are in agreement with the observation of de Martino et al. [1979] on the fate of 'condensed' mitochondria in spermatocytes and spermatids in rat. The distribution pattern of SOx in rat testis is strikingly similar to that of LDH-X which has also been shown to be located within the mitochondrial matrix. Machado de Domenech et al. [1972] reported its association with a special type of mitochondria. Using a histochemical approach, Hintz and Goldberg [1977] reported the presence of LDH-X only from midpachytene stage onwards. With more refined and better techniques, Li et al. [1989] were able to detect LDH-X synthesis at an earlier stage (preleptotene spermatocytes) though at a much lower level. LDH-X activity like SOx increases substantially at midpachytene spermatocyte stage. LDH-X is suggested to mediate the transfer of reducing equivalents between the cytoplasm and mitochondria via a 2-oxoacid-2 redox couple. SOx may also be involved in that shuttle mechanism by regulating the number of sulfhydryl groups and disulfide bonds.

Cell surface markers such as galactosyltransferase and rabbit sperm autoantigen-1 (RSA-1) are other examples whose distribution pattern is similar to SOx [O'Rand et al., 1984; Scully et al., 1987]. The fate of SOx in maturing spermatids seems to be species-dependent. For instance, in rat, immunoreactive material is thrown out of the sperm cell so that sperms free of SOx are released into the lumen. On the other hand, in hamster, the SOx immunoreactivity is seen to be incorporated into the sperm and is localized in the midpiece. This observation in hamster is contrary to the report of Chang and Morton [1975] who observed that caudal sperms are devoid of SOx activity. This difference may be due to (1) shedding of SOx by sperms during their passage to cauda epididymis and/or (2) SOx localized immunologically is functionally an inactive enzyme. The first possibility is ruled out by the observation of Seitz et al. [1988] who reported the presence of SOx within the mitochondria. The second possibility that SOx becomes functionally inactive by the time they reach cauda epididymis, seems plausible. Majority of sulfhydryl groups on caput sperms are oxidized to disulfide bonds as spermatozoa migrate through epididymis [Calvin and Bedford, 1971]. The important role played by SOx in the formation of disulfide bonds on sperm surface is further evident by the observation that disulfide bonding within nucleus and tail structures can occur independently of epididymal environment [Haugaard, 1968]. Haugaard [1968] presumed a spontaneous oxidation of protein thiol groups to disulfide bonds in in vitro studies. We are reporting for the first time the presence of SOx on testicular sperms in hamster. Probably this enzyme is involved in the spontaneous oxidation of protein thiol groups as observed by Haugaard [1968].

Our observations suggest that SOx like LDH-X, RSA-1 and cytochrome c, is another example of differential gene activation associated with a developmental process [O'Rand et al., 1984]. The significance of appearance of SOx during early developments of sperm is not yet known. It has been reported that mitochondrial precursor protein(s) unfold to a protease-sensitive conformation at the surface of mitochondria before being translocated into the mitochondria [Eilers et al., 1988]. We suggest that mitochondrial SOx may help these translocated proteins to stabilize their native tertiary structure.

Secretion of the Coagulating Gland

Transglutaminase
Using SDS-PAGE, four different proteins can be distinguished in the secretion of the coagulating gland. Transglutaminase, the enzyme essential for semen coagulation, could be purified by preparative isoelectric focusing (granulated G-75 gel bed, pH gradient 4–9.5) subsequent to ammonium sulfate precipitation (30–50% saturation). Transglutaminase links polyamines (putrescine, cadaverine, spermine, spermidine) to proteins or forms links between proteins by γ-glutamyl-ε-lysine bridges. On SDS-polyacrylamide gels the protein has a molecular weight of 66 kD; it is glycosylated with mannose residues. The pI of transglutaminase present in crude extracts is 7.7 while it is 8.6 in purified enzyme [Seitz et al., 1987b, 1988]. This change in charge is not related with any significant reduction of molecular weight. For quantitative detection of the enzyme, a precipitation test similar to the Ouchterlony test (immunodiffusion in agarose gels) was developed. SVS II or standardized extract of SVs from sexually inactive animals served as the substrate. Alternatively, a quantification method using ^{14}C-putrescine as a donor and N′N′dimethyl-casein as an acceptor was used. A nephelometric test monitoring light transmission as a function of transglutaminase-treated SVS II solution (at 660 nm) was developed as well [Seitz and Aumüller, 1989; Aumüller and Seitz, 1990a, b]. Analyzing the interaction between SVS II and transglutaminase by SDS-PAGE, we found, contrary to the results of Fawell and Higgins [1987], that among SV proteins nearly exclusively SVS II (and some SVS I) is precipitated by transglutaminase. This precipitate remains insoluble even under most dramatic reducing conditions indicating a stabile covalent linkage of the protein. Only tryptic digestion is capable of leading to destruction of SVS II molecules. An antibody raised in rabbits against secretory transglutaminase from coagulating gland cross-reacted with a protein in rat dorsal prostate. Likewise, cross-reactive with rat transglutaminase was an antibody directed against transglutaminase from guinea pig coagulating gland (80 kD). Spermatozoa taken from various portions of

the epididymis were covered in the head region by a transglutaminase immunoreactive protein. The reaction was strong in the initial portion of the epididymal duct and low in the vas deferens.

Transglutaminase is an enzyme that triggers multifarious biological responses such as receptor stabilization or polymerization of seminal secretory proteins during formation of the copulatory plug in rodents [Williams-Ashman, 1984], and in immune defense mechanisms [James and Hargreave, 1984]. This enzyme is an anchor protein and is released in an apocrine secretion mode from the dorsal prostate and the coagulating gland and therefore may serve as a tool in scrutinizing this special mode of secretion.

References

Abrescia, P.; Lombardi, G.; de Rosa, M.; Quagliozzi, L.; Guardiola, J.; Metafora, S.: Identification and preliminary characterization of a sperm-binding protein in normal human semen. J. Reprod. Fertil. *73:* 71–77 (1985).

Acott, T.S.; Hoskins, D.D.: Bovine sperm forward motility protein: Binding to epididymal spermatozoa. Biol. Reprod. *24:* 234–240 (1981).

Agrawal, Y.; Vanha-Perttula, T.: Effect of secretory particles in bovine seminal vesicle secretion on sperm motility and acrosome reaction. J. Reprod. Fertil. *79:* 409–419 (1987).

Aumüller, G.; Braun, B.E.; Seitz, J.; Müller, T.; Heyns, W.; Krieg, M.: Effects of sexual rest or sexual activity on the structure and function of the ventral prostate of the rat. Anat. Rec. *212:* 345–352 (1985).

Aumüller, G.; Scheit, K.H.: Immunohistochemistry of secretory proteins in bull seminal vesicles. J. Anat. *150:* 43–48 (1987).

Aumüller, G.; Seitz, J.: Immunoelectron microscopic evidence for different compartments in the secretory vacuoles of rat seminal vesicles. Histochem. J. *18:* 15–23 (1986).

Aumüller, G.; Seitz, J.: Protein secretion and secretory processes in male accessory sex glands. Int. Rev. Cytol. *121:* 127–231 (1990a).

Aumüller, G.; Seitz, J.: Transglutaminase immunoreactivity in the male genital tract of the rat. Acta Histochem. (in press, 1990b).

Aumüller, G.; Vesper, M.; Seitz, J.; Kemme, M.; Scheit, K.H.: Binding of a major secretory protein from bull seminal vesicles to bovine spermatozoa. Cell Tissue Res. *252:* 377–384 (1988).

Babcock, D.F.; Singh, J.P.; Lardy, H.A.: Alteration of membrane permeability to calcium ions during maturation of bovine spermatozoa. Dev. Biol. *69:* 85–93 (1979).

Ban, Y.; Wang, M.C.; Watt, K.W.K.; Loor, R.; Chu, M.T.: The proteolytic activity of human prostate-specific antigen. Biochem. Biophys. Res. Commun. *123:* 482–488 (1984).

Bedford, J.M.: Changes in the electrophoretic properties of rabbit spermatozoa during passage through the epididymis. Nature *200:* 1178–1180 (1963).

Brandt, H.; Hoskins, D.D.: A cAMP-dependent phosphorylated motility protein in bovine epididymal sperm. J. Biol. Chem. *255:* 982–987 (1980).

Calvin, H.I.; Bedford, J.M.: Formation of disulfide bonds in the nucleus and accessory structures of mammalian spermatozoa during maturation in the epididymis. J. Reprod. Fertil. *13:* suppl., pp. 65–75 (1971).

Chang, T.S.K.; Morton, B.: Epididymal sulfhydryl oxidase: A sperm-protecting enzyme from the male reproductive tract. Biochem. Biophys. Res. Commun. *66:* 309–315 (1975).

Chang, T.S.K.; Zirkin, B.R.: Distribution of sulfhydryl oxidase activity in the rat and hamster male reproductive tract. Biol. Reprod. *17:* 745–748 (1978).

Comte, M.; Malone, A.; Cox, J.A.: Affinity purification of seminal plasmin and characterization of its interaction with calmodulin. Biochem. J. *240:* 567–573 (1986).

Coronel, C.; San Agustin, J.; Lardy, H.A.: Identification and partial characterization of caltrin-like proteins in the reproductive tract of the guinea pig. Biol. Reprod. *38:* 713–722 (1988).

Daunter, B.; Hill, R.; Hennessey, J.; Machary, E.V.: Seminal plasma biochemistry. I. Preliminary report: A possible mechanism for the liquefaction of human seminal plasma and its relationship to spermatozoal mobility. Andrologia *13:* 131–141 (1980).

de Lamirande, E.; Gagnon, C.: Aprotinin and a seminal plasma factor provide two new tools to study the regulation of sperm motility. J. Submicrosc. Cytol. *15:* 83–87 (1983).

de Martino, C.; Floridi, A.; Marcante, M.L.; Malorni, W.; Scorza Barcellona, P.; Bellocci, M.; Silvestrini, B.: Morphological, histochemical and biochemical studies on germ cell mitochondria of normal rats. Cell Tissue Res. *196:* 1–22 (1979).

Dubé, J.Y.; Frenette, G.; Chapdelaine, P.; Paquin, R.; Tremblay, R.R.: Biochemical characteristics of the proteins secreted by dog prostate, a review. Exp. Biol. *43:* 149–159 (1985).

Dubé, J.Y.; Gandreault, D.; Tremblay, R.R.: The concentration of immunoreactive prostate-specific antigen is not decreased in viscous semen samples. Andrologia *21:* 136–139 (1989).

Dubé, J.Y.; Pelletier, G.; Gagnon, P.; Tremblay, R.R.: Immunohistochemical localization of a prostatic secretory protein of 94 amino acids in normal prostatic tissue, in primary prostatic tumors and in their metastases. J. Urol. *138:* 883–887 (1987).

Eilers, M.; Hwang, S.; Schatz, G.: Unfolding and refolding of a purified precursor protein during import into isolated mitochondria. EMBO J. *7:* 1139–1145 (1988).

Esch, F.S.; Ling, N.C.; Böhlen, P.; Ying, S.Y.; Guillemin, R.: Primary structure of PDC-109, a major protein constituent of bovine seminal plasma. Biochem. Biophys. Res. Commun. *113:* 861–868 (1983).

Evans, R.J.; Herr, J.C.: Immunohistochemical localization of the MHS-5 antigen in principal cells of human seminal vesicle epithelium. Anat. Rec. *214:* 372–377 (1986).

Fawell, S.E.; Higgins, S.J.: Formation of rat copulatory plug: purified seminal vesicle secretory proteins serve as transglutaminase substrates. Mol. Cell. Endocrinol. *53:* 149–152 (1987).

Fawell, S.E.; Pappin, D.J.C.; McDonald, C.J.; Higgins, S.J.: Androgen-regulated proteins of rat seminal vesicle secretion constitute a structurally related family present in the copulatory plug. Mol. Cell. Endocrinol. *45:* 205–213 (1986).

Forestà, C.; Caretto, A.; Indino, M.; Betterle, C.; Scandarelli, C.: Calcitonin in human seminal plasma and its localization on human spermatozoa. Andrologia *18:* 470–473 (1986).

Garde, S.V.; Sheth, A.R.: Immunoperoxidase localization of prostatic inhibin peptide in human, monkey, dog and rat prostates. Anat. Rec. *223:* 181–184 (1989).

Gerhardt, P.G.; Mevag, B.; Tveter, K.J.; Purvis, K.: A systematic study of biochemical differences between the lobes of the rat prostate. Int. J. Androl. *6:* 553–562 (1983).

Gietzen, K.; Galla, H.-J.: Seminal plasmin. An endogenous calmodulin antagonist. Biochem. J. *230:* 277–280 (1985).

Haugaard, N.: Cellular mechanism of oxygen toxicity. Physiol. Rev. *48:* 311–373 (1968).

Herr, J.C.; Spell, D.R.; Conklin, D.J.; Flickinger, C.J.: Electron microscopic immunolocalization of seminal vesicle-specific antigen in human seminal vesicle. Biol. Reprod. *40:* 333–342 (1989).

Herr, J.C.; Summers, T.; McGee, R.S.; Sutherland, M.; Sigman, M.; Evans, R.J.: Characterization of a monoclonal antibody to a conserved epitope on human seminal vesicle-specific peptides: A novel probe/marker system for semen identification. Biol. Reprod. *35:* 773–784 (1986).

Hierowski, M.T.; McDonald, M.W.; Dunn, L.; Sullivan, J.W.: The partial dependency of human prostatic growth factor on steroid hormones in stimulating thymidine incorporation into DNA. J. Urol. *138:* 909–912 (1987).

Hintz, M.; Goldberg, E.: Immunohistochemical localization of LDH-X during spermatogenesis in mouse testes. Dev. Biol. *57:* 375–384 (1977).

Hoskins, D.D.; Johnson, D.J.; Brandt, H.: Evidence for a role for forward motility protein in the epididymal development of sperm motility; in Fawcett, D.W.; Bedford, J.M. (eds): The Spermatozoon, pp. 43–53 (Urban & Schwarzenberg, Munich 1979).

James, K.; Hargreave, T.B.: Immunosuppression by seminal plasma and its possible clinical significance. Immunology *5:* 357–363 (1984).

Kandala, J.C.; Kistler, M.K.; Kistler, W.S.: Methylation of the rat seminal vesicle secretory protein IV gene. Biochem. Biophys. Res. Commun. *126:* 948–953 (1985a).

Kandala, J.C.; Kistler, W.; Kistler, M.K.: Methylation of the rat seminal vesicle secretory protein IV gene. Extensive demethylation occurs in several male accessory sex glands. Biol. Chem. *260:* 15959–15967 (1985b).

Kemme, M.; Madiraju, M.V.V.S.; Krauhs, E.; Zimmer, M.; Scheit, K.H.: The major protein of bull seminal plasma is a secretory product of seminal vesicle. Biochem. Biophys. Acta *884:* 282–290 (1986).

Kirchheiner, J.; Köller, A.; Binder, B.R.: Isolation and characterization of plasminogen activators from hyperplastic and malignant prostate tissue. Biochim. Biophys. Acta *797:* 256–265 (1984).

Koren, E.; Lukac, J.: Mechanism of liquefaction of the human ejaculate. J. Reprod. Fertil. *56:* 493–499 (1979).

Lee, C.; Sensibar, J.A.: Proteins of the rat prostate. II. Synthesis of new proteins in the ventral lobe during castration-induced regression. J. Urol. *138:* 903–908 (1987).

Lewis, R.V.; San Agustin, J.; Kruggel, W.; Lardy, H.: The structure of caltrin, the calcium-transport inhibitor of bovine seminal plasma. Proc. Natl. Acad. Sci. USA *82:* 6490–6491 (1985).

Li, S.-L.; O'Brien, D.A.; Hou, E.W.; Versola, J.; Rockett, D.L.; Eddy, E.M.: Differential activity and synthesis of lactate dehydrogenase isozymes A (muscle), B (heart), and C (testis) in mouse spermatogenic cells. Biol. Reprod. *40:* 173–180 (1989).

Lilja, H.: A kallikrein-like serine protease in prostatic fluid cleaves the predominant seminal vesicle protein. J. Clin. Invest. *76:* 1899–1903 (1985).

Lilja, H.; Abrahamsson, P.-A.: Three predominant proteins secreted by the human prostate gland. Prostate *12:* 29–38 (1988).

Lilja, H.; Abrahamsson, P.-A.; Lundwall, A.: Semenogelin, the predominant protein in human semen. Primary structure and identification of closely related proteins in the male accessory sex glands and on the spermatozoa. J. Biol. Chem. *264:* 1894–1900 (1989).

Lilja, H.; Jeppsson, J.-O.: Amino acid sequence of the predominant basic protein in human seminal plasma. FEBS Lett. *182:* 181–184 (1985).

Lilja, H.; Laurell, C.B.: Liquefaction of coagulated human semen. Scand. J. Clin. Lab. Invest. *44:* 447–452 (1984).

Lilja, H.; Laurell, C.-B.: The predominant protein in seminal coagulate. Scand. J. Clin. Invest. *45:* 635–641 (1985).

Lilja, H.; Oldbring, J.; Rannevik, G.; Laurell, C.-B.: Seminal vesicle-secreted proteins and their reactions during gelation and liquefaction of human semen. J. Clin. Invest. *80:* 281–285 (1987).

Lundquist, F.: Studies on the biochemistry of human semen. 4. Amino acids and proteolytic enzymes. Acta Physiol. Scand. *25:* 178–187 (1952).

Machado de Domenech, E.; Domenech, C.; Aoki, A.; Blanco, A.: Association of testicular lactate dehydrogenase isozyme with a special type of mitochondria. Biol. Reprod. *6:* 136–147 (1972).

Mann, T.: The Biochemistry of Semen and of the Male Reproductive Tract (Methuen, London 1984).

Mann, T.; Lutwak-Mann, C.: Male Reproductive Function and Semen: Themes and Trends in Physiology, Biochemistry and Investigative Andrology (Springer, Berlin 1981).

McGee, R.S.; Herr, J.C.: Human seminal vesicle-specific antigen during semen liquefaction. Biol. Reprod. *37:* 431–439 (1987).

Meizel, S.: Molecules that initiate or help stimulate the acrosome reaction by their interaction with the mammalian sperm surface. Am. J. Anat. *174:* 285–302 (1985).

Minakata, K.; Asano, M.: Acidic cysteine proteinase inhibitor in seminal plasma. Biol. Chem. Hoppe-Seyler *366:* 15–18 (1985).

O'Rand, M.G.; Irons, G.P.; Porter, J.P.: Monoclonal antibodies to rabbit sperm autoantigen. I. Inhibition of in vitro fertilization and localization on the egg. Biol. Reprod. *30:* 721–729 (1984).

Ostrowski, M.C.; Kistler, W.S.: Properties of a flavoprotein sulfhydryloxidase from rat seminal vesicle secretion. Biochemistry *19:* 2639–2645 (1980).

Ostrowski, M.C.; Kistler, M.K.; Kistler, W.S.: Purification and cell-free synthesis of a major protein from rat seminal vesicle secretion. J. Biol. Chem. *254:* 383–390 (1979a).

Ostrowski, M.C.; Kistler, W.S.; Williams-Ashman, H.G.: A flavoprotein responsible for the intense sulfhydryloxidase activity of rat seminal vesicle secretion. Biochem. Biophys. Res. Commun. *87:* 171–176 (1979b).

Polak, B.; Daunter, B.: Seminal plasma biochemistry. IV. Enzymes involved in the liquefaction of human seminal plasma. Int. J. Androl. *12:* 187–194 (1989).

Reddy, E.S.P.; Bhargava, P.M.: Seminal plasmin, an antimicrobial protein from bovine seminal plasma which acts in *E. coli* by specific inhibition of rRNA synthesis. Nature *279:* 725–728 (1979).

Reid, W.A.; Vongsorasak, J.; Valler, S.M.J.; Kay, J.: Identification of the acid proteinase in human seminal fluid as a gastricsin originating in the prostate. Cell Tissue Res. *236:* 597–600 (1984).

Rifkin, J.M.; Olson, G.E.: Characterization of maturation-dependent extrinsic proteins of the rat sperm surface. J. Cell Biol. *100:* 1582–1591 (1985).

Rufo, G.A.; Singh, J.P.; Babcock, D.F.; Lardy, H.A.: Purification and characterization of a calcium transport inhibitor protein from bovine seminal plasma. J. Biol. Chem. *257:* 4627–4632 (1982).

San Agustin, J.T.; Hugues, P.; Lardy, H.A.: Properties and function of caltrin, the calcium-transport inhibitor of bull seminal plasma. FASEB J. *1:* 60–66 (1987).

Scheit, K.H.: The major basic proteins of bull seminal vesicle secretion. Biol. Chem. Hoppe-Seyler *367:* 229–233 (1986).

Scheit, K.H.; Reddy, E.S.P.; Bhargava, P.M.: Seminal plasmin is a potent inhibitor of *Escherichia coli* DNA polymerase in vitro. Nature *279:* 728–732 (1979).

Scheit, K.H.; Shivaji, S.; Bhargava, P.M.: Seminal plasmin, an antimicrobial protein from bull seminal plasma, inhibits growth and synthesis of nucleic acids and proteins in *S. cerevisiae*. J. Biochem. *97:* 463–471 (1985).

Scully, N.F.; Shaper, J.H.; Shur, B.D.: Spatial and temporal expression of cell surface galactosyltransferase during mouse spermatogenesis and epididymal maturation. Dev. Biol. *124:* 111–124 (1987).

Seidah, N.G.; Ramasharma, M.R.; Sairam, M.R.; Chretien, M.: Partial amino acid sequence of a human seminal plasma peptide with inhibin-like activity. FEBS Lett. *167:* 98–102 (1984).

Seitz, J.; Aumüller, G.: Secretory proteins from male accessory sex glands: Isolation, biochemical and functional characterization; in Holstein, F.; Voigt, K.D.; Grässlin, D. (eds): Reproductive Biology and Medicine. A, pp. 112–118 (Diesbach, Berlin 1989).

Seitz, J.; Keppler, C.; Aumüller, G.: Sulfhydryloxidase – ein FAD-abhängiges, Sauerstoffverbrauchendes Enzym im Sekret der Rattenbläschendrüse. Anat. Anz. *164:* suppl., pp. 483–484 (1989).

Seitz, J.; Keppler, C.; Bergmann, M.; Aumüller, G.: Thiol oxidase: a new mitochondrial enzyme. Characterization and immunocytochemical localization (abstract). Eur. J. Cell Biol. *46:* 69 (1988).

Seitz, J.; Keppler, C.; Fahimi, H.D.; Völkl, A.: Staining gels and blots for native oxidases by means of the cerium method. Anal. Biochem. (submitted, 1990).

Seitz, J.; Plog, A.; Keppler, C.; Dreher, M.; Aumüller, G.: Transglutaminase, ein sekretorisches Protein der Koagulationsdrüse und der dorsalen Prostata der Ratte. Verh. Anat. Ges. *81:* 837–839 (1987b).

Shivaji, S.; Bhargava, P.M.; Scheit, K.H.: Immunological identification of tissue extracts of sex glands of bull. Biol. Reprod. *30:* 1237–1241 (1984).

Sidhu, K.S.; Guraya, S.S.: Calmodulin-like protein in the buffalo *(Bubalus bubalis)* seminal plasma and its effect on sperm Ca^{2+}, Mg^{2+}-ATPase. Int. J. Androl. *12:* 148–154 (1989).

Sitaram, N.; Kumari, K.V.; Bhargava, P.M.: Seminal plasmin and caltrin are the same protein. FEBS Lett. *201:* 233–236 (1986).

Sjöberg, H.E.; Arver, S.; Bucht, E.: High concentration of immunoreactive calcitonin of prostatic origin in human semen. Acta Physiol. Scand. *110:* 101–102 (1980).

Story, M.T.; Esch, F.; Shimasaki, S.; Sasse, J.; Jacobs, S.C.; Lawson, R.K.: Amino-terminal sequence of a large form of basic fibroblast growth factor isolated from human benign prostatic hyperplastic tissue. Biochem. Biophys. Res. Commun. *142:* 702–709 (1987).

Tash, J.S.; Krinks, M.; Patel, J.; Means, L.; Klee, C.B.; Means, A.R.: Identification characterization and functional correlation of calmodulin-dependent protein phosphatase in sperm. J. Cell Biol. *106:* 1625–1633 (1988).

Tash, J.S.; Means, A.R.: Regulation of protein phosphorylation and mobility of sperm by cyclic adenosine monophosphate and calcium. Biol. Reprod. *26:* 745–763 (1982).

Tauber, P.; Propping, P.; Schumacher, G.B.F.; Zaneveld, L.J.D.: Biochemical aspects of the coagulation and liquefaction of human semen. J. Androl. *1:* 281–288 (1980).

Tauber, P.F.; Zaneveld, L.J.D.; Propping, D.; Schumacher, G.F.P.: Components of human split ejaculates. II. Enzymes and proteinase inhibitors. J. Reprod. Fertil. *46:* 165–171 (1976).

Tenniswood, M.: Role of epithelial-stromal interactions in the control of gene expression in the prostate: A hypothesis. Prostate *9:* 375–385 (1986).

Theil, R.; Scheit, K.H.: Amino acid sequence of seminal plasmin, an antimicrobial protein from bull semen. EMBO J. *2:* 1159–1163 (1983).

Wagner, C.L.; Kistler, W.S.: Analysis of the major large polypeptides of rat seminal vesicle secretion: SVS I, II and III. Biol. Reprod. *36:* 501–510 (1987).

Wang, M.C.; Valenzuela, L.A.; Murphy, G.P.; Chu, T.M.: Purification of a human prostate specific antigen. Invest. Urol. *17:* 159–163 (1979).

Weiss, G.: Relaxin in the male. Minireview. Biol. Reprod. *40:* 197–200 (1989).

Wichmann, L.; Vaalasti, A.; Vaalasti, T.; Tuohimaa, P.: Localization of lactoferrin in the male reproductive tract. Int. J. Androl. *12:* 179–186 (1989).

Williams, L.; McDonald, C.; Higgins, S.: Sequence organization of rat seminal vesicle F protein gene: Location of transcriptional start point and sequence comparison with six other androgen-related genes. Nucleic Acids Res. *13:* 659–672 (1985).

Williams-Ashman, H.G.: Transglutaminases and the clotting of mammalian seminal fluids. Mol. Cell. Biochem. *58:* 51–61 (1984).

Dr. Gerhard Aumüller, Abteilung für Anatomie und Zellbiologie,
Philipps-Universität, D-W–3550 Marburg (FRG)

Colpi GM, Pozza D (eds): Diagnosing Male Infertility.
Prog Reprod Biol Med. Basel, Karger, 1992, vol 15, pp 49–56

The Evaluation of Markers of Prostatic Function

H. von der Kammer[a], *K.H. Scheit*[a], *W. Weidner*[b], *T.G. Cooper*[c, 1]

[a] Max-Planck-Institut für biophysikalische Chemie, Göttingen;
[b] Department of Urology, Georg August University, Göttingen, and
[c] Institute of Reproductive Medicine, University of Münster, FRG

The assessment of a range of biochemical markers in human semen is recommended to reveal disturbances of genital tract organs [Cooper et al., 1990]. Of prostatic markers, the concentration of citrate and activity of acid phosphatase are frequently used since a good correlation exists between their concentrations in semen and immunoassayable PAP is also well correlated with its enzyme activity [Wetterauer, 1986]. As demonstrated by Lilja and Abrahamsson [1988], three predominant proteins are secreted by the human prostate gland: prostatic-secreted acid phosphatase (PAP) [McCarthy et al., 1983], prostate-specific antigen (PSA) [Wang et al., 1979; Lilja, 1985] and a 10.7-kD protein called prostatic secretory protein of 94 amino acids (PSP94). The amino acid sequence of PSA has been reported by Watt et al. [1986] and Schaller et al. [1987]. The amino acid sequence of PSP94 was independently determined by Johansson et al. [1984], Seidah et al. [1984] as well as Akiyama et al. [1985]. Recently the cDNAs specific for these predominantly prostate-derived proteins were characterized and sequenced: the PAP-specific cDNA [Vihko et al., 1988], the PSA-specific cDNA [Lundwall and Lilja, 1987] as well as PSP94-specific cDNA [Mbikay et al., 1987]. The immunohistochemical distribution of the secretory proteins of the prostate in the parenchyma of prostate

[1] We thank Prof. E. Nieschlag for support and encouragement.

glands was investigated by Abrahamsson et al. [1988]. The localization of PSP94 in epithelial cells of the prostate was demonstrated both by in situ hybridization employing PSP94-specific cDNA as well as by immunohistochemistry [Brar et al., 1988]. The present study compares the prostatic secretion of these proteins with that of citrate, an established marker of prostatic function, fructose as an indicator of seminal vesicle secretion and glucosidase as a marker for the epididymal status of the genital organ [Cooper et al., 1990].

Materials and Methods

Semen. Semen was obtained by masturbation from men attending the Outpatient Department for Prostatitis and Urological Andrology, Giessen. The samples were taken from the same collective of patients previously reported [Cooper et al., 1988; 1990] as suffering from (a) epididymitis, (b) chronic prostatitis and (c) chronic prostato-urethritis (adnexitis). Normozoospermic semen samples from men who recently attended the clinic because of a barren marriage were taken as controls since there was no evidence of inflammation of the genital tract. After liquefaction, semen was centrifuged at 2,000 g for 10 min and frozen at $-70\,°C$ for 3–4 years. Samples were thawed and well mixed between assays.

Assays. PSA and PAP were measured by commercial ELISA (Hybritech, CIS) and citrate, fructose, glucosidase by spectrophotometric methods reported previously [Cooper et al., 1990]. PSP94 was determined by an ELISA described recently [von der Kammer et al., 1990]. All determinations for PAP, PSA and PSP94 were carried out in duplicate. The intra-assay variation in ELISA measurements was $\leq 8\%$. The output per organ of the markers was obtained by multiplying its concentration in semen with the volume.

Statistical Analysis. Because the data obtained from the respective semen collectives did not follow a standard distribution, statistically significant differences between groups were detected by calculating the median at different confidence levels [Sachs, 1988]. Confidence ranges of median values of disease groups not overlapping with those of the normal collective were considered to be significantly different. Calculations were performed by means of the program StatView on a Macintosh Plus computer.

Results

Employing the experimental data measured for the normal noninflammatory semen samples (group N) as a reference, a correlation of the parameters was attempted. The results depicted in table 1 clearly indicate a sig-

Table 1. Matrix of correlation coefficients of various parameters for 32 samples of normal semen

	PSP94	PAP	PSA	Gluco-sidase	Fructose	Citrate
PSP94, mg/ejaculate	1.000	0.686*	0.696*	0.225	0.048	0.823*
PAP, mg/ejaculate	0.686*	1.000	0.447	−0.000	0.049	0.853*
PSA, mg/ejaculate	0.696*	0.447	1.000	0.259	0.352	0.642*
Glucosidase, mU/ejaculate	0.225	−0.000	0.259	1.000	0.408	0.214
Fructose, μmol/ejaculate	0.048	0.049	0.352	0.408	1.000	0.084
Citrate, μmol/ejaculate	0.823*	0.853*	0.642*	0.214	0.048	1.000

Correlation coefficients ≥ 0.60 are marked by an asterisk.

nificantly positive correlation between the protein markers PAP, PSA, PSP94 as well as between these markers and citrate; inflammation of genital tract tissue did not affect this correlation. No correlation with fructose or glucosidase was detected.

In a first attempt to illustrate differences between the parameter values of the semen collectives, the percentile plots of the data were inspected (fig. 1a–f). A clear-cut picture is derived from a comparison of the percentile plots of fructose and PSA values of groups A, N, E and P; the median as well as the 90th percentile distribution of values are apparently not significantly different. In the percentile plot, glucosidase appears uniquely related to an inflammatory condition of the epididymis: almost 75% of all glucosidase values of group E were found to be lower than the 25th percentile of the normal values. In contrast, the distribution of glucosidase values of groups A and P were like those of the normal collective.

A perplexing property of PAP secretion from individuals suffering from inflammatory processes is indicated by the percentile plots of groups A, E and P. The median values of all are apparently elevated when compared to the normal group. However, the distribution of values between the 25th and 75th percentile is rather broad and overlaps with that of group N. Close inspection of the distribution of citrate values of group P in the percentile plot disclosed that more than 25% of the citrate values were found to be lower than the lowest citrate values of group N. On the other hand, the median levels of citrate values for all groups did not differ significantly.

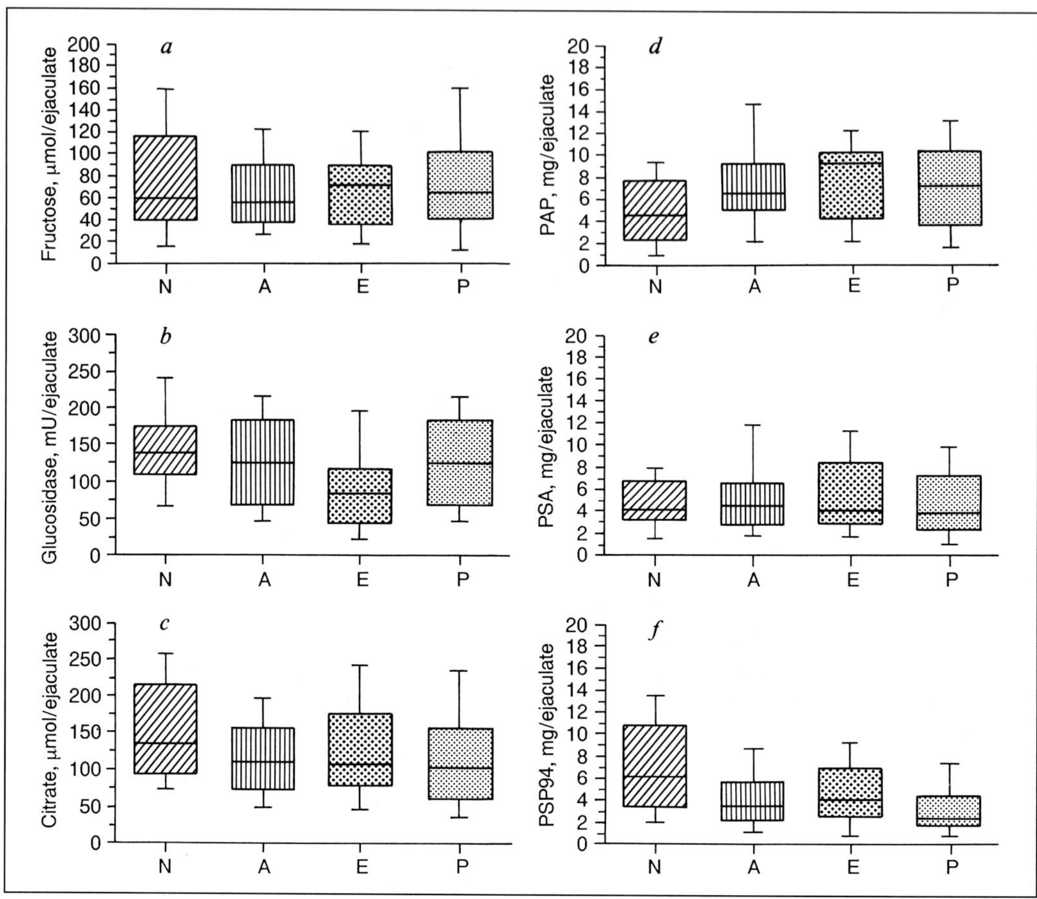

Fig. 1. Schematic representation of percentiles of semen parameters. Abbreviation of semen collectives: N = normozoospermic, noninflammatory; A = adnexitis; E = acute epididymitis; P = chronic prostatitis. The lines of individual boxes represent, from top to bottom, the 90th, 75th, 50th, 25th and 10th percentiles.

The percentile plot of PSP94 values illustrates differences between values of groups A and P with regard to the normal collective. In the case of group A, 75% of all values are below the median of normal values and the distribution of values between the 25th as well as the 75th percentile is narrow. Moreover, in group P, the median level is well below the 25th percentile of the values for group N.

Table 2. Amounts of fructose, citrate, glucosidase and prostatic secretory proteins in seminal plasma in various inflammatory conditions

Parameter	Diagnosis	n	Range	Median	95% confidence range	99% confidence range
Fructose	N	32	2–273	59.7	40.0–98.0	38.5–111.6
μmol/ejaculate	A	39	6–184	56.1	39.6–76.4	39.0–79.75
	E	25	11–122	70.8	39.6–80.4	33.6–87.6
	P	62	4–241	64.0	55.2–72.3	46.2–73.0
Citrate	N	32	69–476.7	133.8	98.0–178.5	93.0–206.0
μmol/ejaculate	A	38	32–371	110.6	76.8–146.4	72.5–151.2
	E	23	32.4–244.2	106.0	86.4–144.0	69.6–183.0
	P	62	11.4–360	102.05	69.6–125.5	64.2–131.6
Glucosidase	N	32	42–299.5	138.75	126.60–166.2	115.5–171.0
mU/ejaculate	A	42	5.8–308.4	114.05	91.8–147.4	88.5–158.5
	E	26	11.7–333.6	84.0	42.0–113.4*	41.0–116.8*
	P	43	11.4–374.1	124.2	101.75–146.0	86.5–165.6
PAP	N	31	0.28–27.49	4.580	2.948–6.465	1.998–7.080
mg/ejaculate	A	40	0.32–16.05	6.525	5.160–8.250	5.060–8.420
	E	22	0.44–13.0	9.295	4.200–10.120	2.940–10.320
	P	62	0.44–21.51	7.200	4.860–8.550	4.470–8.820
PSA	N	32	0.16–10.98	4.175	3.35–5.76	3.12–6.45
mg/ejaculate	A	39	0.66–16.6	4.10	3.00–5.28	2.75–5.88
	E	23	1.26–16.88	4.050	3.32–7.00	2.68–8.72
	P	61	0.28–20.25	3.875	2.590–5.430	2.34–6.04
PSP94	N	32	1.16–20.3	6.070	4.120–9.000	3.630–9.180
mg/ejaculate	A	41	0.54–16.2	3.452	2.648–4.475	2.252–4.794
	E	24	0.44–13.38	3.998	2.620–6.810	1.892–6.990
	P	61	0.32–17.57	2.574	1.864–3.064*	1.832–3.108*

Confidence ranges of median values which show no or insignificant overlap with that of the normal group are marked by an asterisk. For abbreviations of diagnosis see figure 1.

To evaluate the statistical significance of the differences displayed in the percentile presentation, confidence ranges for the median levels of the parameters for the different groups were calculated. The surprising result (table 2) is that the citrate values for group P, in comparison to group N, even at the lower confidence range of 95%, were statistically insignificant. However, the parameters glucosidase as well as PSP94 both at the 95% and 99% confidence range appeared to be significantly related to the inflammatory condition epididymitis and prostatitis, respectively. Various ratios of marker concentration were calculated to see if better discrimination between the groups of inflammatory diseases could be achieved, but improved differentiation was not obtained.

Discussion

Measurement of a range of markers in semen as indices of accessory gland function provides additional data that can help to confirm a clinical diagnosis [Cooper et al., 1990]. The inflammatory conditions described here did not appear to influence the seminal vesicles, since fructose secretion was not different in the disease groups from that in the controls. As expected, total glucosidase activity was significantly lower in epididymitis than in the control group. Although the secretion of the accepted prostatic marker citrate was apparently reduced in comparison with controls in prostatitis, as anticipated, and further also in epididymitis or adnexitis, as previously demonstrated [Cooper et al., 1990], the differences were however statistically not significant.

Of the new markers examined here, only the PSP94 suffered a reduced secretion in cases of prostatitis. Of interest is the observation that the secretion of PSA was not influenced by the inflammatory condition. Clearly, different mechanisms of synthesis and for secretion of the prostatic proteins are evident here. The opposite trend in the secretion of PAP and citrate compared to the normal group requires comment. It is usual for this enzyme activity to be well correlated to the ejaculate content of citrate [Wetterauer, 1986]. Since in this study citrate was reduced and PAP was increased, relative to the control semen group, there would appear here to be a dichotomy between the activity of the enzyme and the amount of enzyme protein detected by the antibody in the ELISA. However, as indicated above, if corresponding data pairs are compared, the known correlation between secretion of citrate and PAP remains valid for the groups of semen samples from patients suffering from inflammation of the genital tract. As PSP94 secretion was also decreased in the conditions of adnexitis and epididymitis, secretion of PSP94 would appear to be more sensitive to inflammation than citrate. We therefore advocate measurement of PSP94 in human semen as a useful marker of prostatic function.

References

Abrahamsson, P.A.; Lilja, H.; Falkmer, S.; Wadström, L.B.: Immunohistochemical distribution of the three predominant secretory proteins in the parenchyma of hyperplastic and neoplastic prostate glands. Prostate *12:* 39–46 (1988).

Akiyama, K.; Yoshioka, Y.; Schmid, K., et al.: The amino acid sequence of human β-microseminoprotein. Biochim. Biophys. Acta *829:* 288–294 (1985).

Brar, A.; Mbikay, M.; Sirois, F.; Fournier, S.; Seidah, N.G.; Chretien, M.: Localization of the human prostatic secretory protein PSP94 and its mRNA in the epithelial cells of the prostate. J. Androl. *9:* 253–260 (1988).

Cooper, T.G.; Weidner, W.; Nieschlag, E.: The influence of inflammation of the human male genital tract on the secretion of the seminal markers α-glucosidase, glycerophosphocholine, carnitine, fructose and citric acid. Int. J. Androl. *13:* 329–336 (1990).

Cooper, T.G.; Yeung, C.H.; Nashan, D.; Nieschlag, E.: Epididymal markers in human infertility. J. Androl. *9:* 91–101 (1988).

Johansson, J.; Sheth, A.; Cederlund, E.; Jörnvall, H.: Analysis of an inhibin preparation reveals apparent identity between a peptide within inhibin-like activity and a sperm-coating antigen. FEBS Lett. *176:* 21–26 (1984).

Kammer, H. von der; Krauhs, E.; Aumüller, G.; Scheit, K.H.: Characterization of a monoclonal antibody specific for prostatic secretory protein of 94 amino acids (PSP94) and development of a two-site binding enzyme immunoassay for PSP94. Clin. Chim. Acta (in press, 1990).

Lilja, H.: A kallikrein-like serine protease in prostatic fluid cleaves the predominant seminal vesicle protein. Clin. Invest. *76:* 1899–1903 (1985).

Lilja, H.; Abrahamsson, P.A.: Three predominant proteins secreted by the human prostate gland. Prostate *12:* 29–38 (1988).

Lundwall, A.; Lilja, H.: Molecular cloning of human prostate-specific antigen cDNA. FEBS Lett. *214:* 317–322 (1987).

Mbikay, M.; Nolet, S.; Fournier, S.; Benjannet, S.; Chapdelaine, P.; Paraids, G.; Dubé, J.Y.; Tremblay, R.; Lazure, C.; Seidah, N.G.; Chretien, M.: Molecular cloning and sequence of the cDNA for a 94-amino acid seminal plasma protein secreted by the human prostate. DNA *6:* 23–29 (1987).

McCarthy, R.C.; Jakubowski, H.V.; Markowitz, H.: Human prostatic acid phosphatase: Purification, characterization and optimization conditions for radioimmunoassay. Clin. Chim. Acta *132:* 287–299 (1983).

Sachs, L.: Statistische Methoden: Planung und Auswertung, p. 60 (Springer, Berlin 1988).

Schaller, J.; Akiyama, K.; Tsuda, R.; Marti, T.; Rickli, E.E.: Isolation, characterization and amino-acid sequence of gamma-seminoprotein, a glycoprotein from human seminal plasma. Eur. J. Biochem. *170:* 111–120 (1987).

Seidah, N.G.; Arbatti, N.J.; Rochemont, J.; Sheth, A.; Chretien, M.: Complete amino acid sequence of human seminal plasma β-inhibin. FEBS Lett. *175:* 349–355 (1984).

Vihko, P.; Virkkunnen, P.; Henttu, P.; Roiko, K.; Solin, T.; Huhtala, M.L.: Molecular cloning and sequence analysis of cDNA encoding human prostatic acid phosphatase. FEBS Lett. *236:* 275–281 (1988).

Wang, M.C.; Valenzuela, L.A.; Murphy, G.P.; Chu, T.M.: Purification of a human prostate-specific antigen. Invest. Urol. *17:* 159–163 (1979).

Watt, K.W.K.; Lee, P.J.L.; M'Timkulu, T.M.; Chan, W.P.; Loor, R.: Human prostate-specific antigen: Structural and functional similarity with serine proteases. Proc. Natl. Acad. Sci. USA *83:* 3166–3170 (1986).

Wetterauer, U.: Recommended biochemical parameters for routine semen analysis. Urol. Res. *14:* 241–246 (1986).

H. von der Kammer, Max-Planck-Institut für biophysikalische Chemie, D-W–3400 Göttingen (FRG)

Colpi GM, Pozza D (eds): Diagnosing Male Infertility.
Prog Reprod Biol Med. Basel, Karger, 1992, vol 15, pp 57–61

Measurement of Acrosin Activity in a Group of Patients with Unexplained Infertility

Ashok Agarwal, Kevin R. Loughlin

Division of Urology, Department of Surgery, Brigham and Women's Hospital,
Harvard Medical School, Boston, Mass., USA

The mechanisms of successful fertilization remain poorly understood and as a result many men who are treated for infertility have no identifiable cause for their problem. Acrosin, a sperm-specific acrosomal proteinase, has an essential role in the fertilization process. However, until recently little attention was given to the role of acrosin as part of the explanation for some cases of male infertility. This may have been due, in part, to the fact that the available assays for acrosin [1–3] were arduous and difficult to perform. The recent development of a more simplified reliable technique for measurement of acrosin activity [4] made it attractive to study acrosin levels in men being treated for idiopathic infertility.

Materials and Methods

The methods developed and popularized by Kennedy et al. [4] to assay acrosin activity were used in our study and appear below.

Reagents

Benzamidine hydrochloride, Ficoll (type 400), N-α-benzoyl-*DL*-arginine *p*-nitroanilide hydrochloride (BAPNA), dimethylsulfoxide (DMSO), Hepes (N-2-hydroxyethylpiperazine-N-2-ethanesulfonic acid) and Triton X-100 were obtained from Sigma Chemical Co. (St. Louis, Mo.). Semen samples were obtained from apparently normal healthy men (n = 4) and from suspected subfertile men (n = 10).

Solutions

Solution A: Ficoll: 11% Ficoll in 0.12 M NaCl, 0.025 M Hepes buffer at pH 7.4–7.6. The solution is prepared by dissolving 0.70 g NaCl, 0.60 g Hepes and 11.0 g Ficoll in approximately 90 ml of distilled, deionized water. The solution was adjusted to pH 7.4–7.6 (with HCl or NaOH as required) and was then adjusted to a final volume of 100 ml with water. Sodium azide (0.1%) was added as a preservative in the Ficoll solution for long-term storage in the refrigerator.

Solution B: Detergent buffer: 0.01% Triton X-100 in 0.055 M Hepes, 0.055 M NaCl at pH 8.0. The solution is prepared by dissolving 1.31 g of Hepes, 0.32 g of NaCl and 1 ml of a 1% Triton X-100 stock solution (1 ml Triton in 99 ml water) in 95 ml of distilled, deionized water. The solution is adjusted to pH 8.0 (with 1 N NaOH as required) and then adjusted to a final volume of 100 ml with distilled deionized water. Sodium azide (0.1%) was added as preservative in the detergent buffer for long-term storage in the refrigerator.

Solution C: Benzamidine: 500 mM in water. The solution was prepared by dissolving 87.3 g benzamidine-HCl in 1 liter distilled, deionized water. The solution was stored in the refrigerator for up to 2 weeks.

Solution D: Substrate: 23 mM BAPNA in DMSO. The solution was prepared by dissolving 25 mg BAPNA in 2.5 ml DMSO. The solution was prepared fresh on the day of the assay.

Solution E: Substrate-detergent mixture: 22.5 ml of the detergent buffer (solution B) was mixed thoroughly with 2.5 ml of the BAPNA/DMSO substrate solution (solution D) in a 50-ml culture flask.

Assay Procedure

(1) The ejaculate was allowed to liquefy for 15–30 min. Sperm concentration was measured on a Cell-Soft 2000 computerized semen analyzer.

(2) For each ejaculate, a volume was calculated that will contain 2–10 million spermatozoa/ml. No more than 250 µl was layered over Ficoll except in cases of severe oligozoospermia.

(3) One control and two tests were run simultaneously for each ejaculate. Aliquots (0.5 ml) of the Ficoll (solution A) were pipetted into 15-ml plastic, conical centrifuge tubes. The ejaculate was thoroughly mixed and the calculated volume of the ejaculate was layered over the Ficoll in each tube (including the control tube) preferably with a pipettor (Pipetman; Rainin Instrument Co., Woburn, Mass.).

(4) The tubes were centrifuged at 1,000 g for 30 min at room temperature.

(5) After centrifugation, the seminal plasma and Ficoll supernatant was removed by suction through a thin-stemmed (Pasteur) pipette. About 0.1 ml of the Ficoll was left behind in the tube.

(6) 100 µl of the benzamidine solution (solution C) was immediately added to the control tube and the contents of the tube were mixed thoroughly by vortexing.

(7) To each tube, including the control, 1 ml of the substrate-detergent mixture (solution E) was added and the contents of the tube were mixed thoroughly.

(8) The tubes were incubated at 22–24 °C for exactly 3 h after the addition of the assay solution. The contents of the tubes were mixed once every hour during the incubation period.

(9) After 3 h of incubation, 100 µl of the benzamidine solution (solution C) was added to all the tubes except the control.

(10) All the tubes were centrifuged at 1,000 *g* for 30 min and the supernatant solutions were collected separately.

(11) A LKB spectrophotometer (Ultrospec II) was adjusted to an absorbance reading of 0.0 at 410 nm with the substrate-detergent mixture (solution E). Subsequently, the absorbance of each supernatant solution was recorded at 410 nm. The acrosin activity of the ejaculate was calculated according to the following formula:

$$\mu IU \text{ acrosin/million sperm} = \frac{(\text{mean OD test} - \text{OD control}) \times 10}{1{,}485 \times \text{number of sperm (in millions layered over Ficoll)}}$$

Results

The results of routine semen analysis parameters and acrosin activity in 10 men being evaluated for idiopathic infertility and 4 fertile controls are listed in table 1.

Table 1. Semen analysis and acrosin activity

Patient No.	Sperm concentration 10^6/ml	Motility %	Sperm layered over Ficoll 10^6/ml	Ejaculate volume, ml	Acrosin value $\mu IU/10^6$ sperm
Fertility controls					
1	80	74	10/0.15	1.0	40
2	63	72	15/0.25	1.5	26
3	72	55	8.5/0.1	3.1	28
4	240	88	12/0.05	3.5	36
Idiopathic infertile patients					
5	188	93	9.4/0.05	1.8	29.7
6	22	49	11/0.5	3.0	17.7
7	193	67	19/0.1	3.0	25.0
8	41	23	12/0.3	1.2	8.0
9	50	52	12/0.25	0.9	9.6
10	21	23	5/0.25	2.0	5.6
11	20	28	10/0.5	2.5	8.7
12	15	38	7.4/0.5	6.5	10.6
13	65	59	13/0.2	2.5	10.9
14	19	61	9.5/0.5	3.5	21.4

Discussion

The act of fertilization is a complex process that involves multiple steps, some of which are only beginning to be understood. Acrosin is a sperm-specific acrosomal proteinase which has an essential role in the fertilization process. Acrosin can be found on spermatozoa in both an active (nonzymogen) and inactive (zymogen) form. The inactive form is termed proacrosin and represents 90% of the acrosin on human spermatozoa [3]. From a clinical standpoint, it appears most useful to measure active acrosin rather than total acrosin content. The methodology used in our study, which was introduced by Kennedy et al. [4], more accurately measures active acrosin levels than methods which employ acrosin antibodies or RIA and ELISA which reflect total acrosin levels. Kennedy et al. [4] have established normal and abnormal acrosin ranges as follows: Normal ≥ 25 $\mu IU/ml/10^6$ spermatozoa; indeterminate 14–25 $\mu IU/ml/10^6$ spermatozoa; infertile ≤ 14 $\mu IU/ml/10^6$ spermatozoa.

All 4 of our fertile donors scored in the fertile range (25 $\mu IU/ml/10^6$ spermatozoa) as defined by Kennedy et al. [4]. However, only 2 of 10 of our infertile patients demonstrated acrosin levels in the normal range, whereas 6 out of 10 of the infertile patients exhibited acrosin levels in the subfertile range (≤ 14 $\mu IU/ml/10^6$ spermatozoa). The remaining 2 patients with clinical infertility had acrosin levels in the indeterminate range. Of particular note are 3 patients (Nos. 6, 9, 13) who had essentially normal semen parameters, but exhibited low acrosin levels.

Increasing attention is being paid to acrosin activity and its role in fertilization. The presence of acrosin appears to be critical to the process of fertilization. Spermatozoa from apparently normal, asymptomatic men possess approximately three times the acrosin as those from subfertile men [5, 6]. Cryopreservation, which is known to decrease sperm-fertilizing ability, has been associated with decreased acrosin levels [7]. Finally, Florke-Gerloff et al. [8] and Jeyendran et al. [9] have reported that acrosomeless spermatozoa with almost undetectable levels of acrosin are infertile even if they are motile and possess an intact plasma membrane.

Conclusion

Acrosin appears to play a crucial role in the fertilization process. Patients with normal routine semen parameters may have decreased acrosin activity which may explain the cause of their infertility. Further studies of factors influencing acrosin activity are indicated if we are to achieve a greater understanding of the factors contributing to successful fertilization.

References

1 Fiscor G, Ginsberg LC, Oldford GM, Snoke RE, Becker RW: Gelatin substrate film technique for detection of acrosin in single mammalian sperm. Fertil Steril 1983;39: 548–552.

2 Fritz H, Schleuning WD, Schiessler H, Schill WB, Windt V, Winkler G: Boar, bull and human sperm acrosin isolation, properties and biological aspects; in Reich E, Ritkin DB, Shaw E (eds): Proteases and Biological Control. Cold Spring Harbor, Cold Spring Harbor Laboratory, 1975, pp 715–735.

3 Goodpasture JC, Polakoski KL, Zaneveld LJD: Acrosin, proacrosin and acrosin inhibitor of human spermatozoa: Extraction, quantitation and stability. J Androl 1980;1:16–27.

4 Kennedy WP, Kaminski JM, Van der Ven HH, Jeyendran RS, Reid DS, Blackwell J, Bielfeld P, Zaneveld LJD: A simple clinical assay to evaluate the acrosin activity of human spermatozoa. J Androl 1989;10:221–231.

5 Goodpasture JC, Zavos PM, Cohen MR, Zanefeld LJD: Relationship of human sperm acrosin and proacrosin to semen parameters. J Androl 1982;3:151–156.

6 Mohsenian M, Syner FN, Moghissi KS: A study of sperm acrosin in patients with unexplained infertility. Fertil Steril 1982;37:223–229.

7 Mack SR, Zaneveld LJD: Comparative activation studies with extracted and purified human proacrosin. Comp Biochem Physiol 1986;83:537–543.

8 Florke-Gerloff S, Topter-Peterson E, Muller-Esterl W, Mansouri A, Schatz R, Schirren C, Schill W, Engel W: Biochemical and genetic investigations of round-headed spermatozoa in infertile men including two brothers and their father. Andrologia 1984;16:187–202.

9 Jeyendran RS, Van der Ven HH, Kennedy W, Heath E, Perez-Palaez M, Sobrero AJ, Zaneveld LJD: Acrosomeless sperm: A cause of primary male infertility. Andrologia 1985;17:31–36.

Dr. Ashok Agarwal, Division of Urology, Department of Surgery,
Brigham and Women's Hospital, 45 Francis Street, Boston, MA 02115 (USA)

Colpi GM, Pozza D (eds): Diagnosing Male Infertility.
Prog Reprod Biol Med. Basel, Karger, 1992, vol 15, pp 62–65

Effect of Sperm Washing with Theophylline on Human Spermatozoa Fertilizing Ability

Ashok Agarwal, Kevin R. Loughlin

Division of Urology, Department of Surgery, Brigham and Women's Hospital,
Harvard Medical School, Boston, Mass., USA

Various substances have been utilized to improve sperm function. A list of substances that have been used and their proposed mechanisms of action is listed in table 1. Our study was designed to examine the effects of theophylline, an agent known to increase intracellular levels of cAMP, on the fertilizing capacity of suspected infertile men and fertile controls as measured by the hamster egg penetration test.

Materials and Methods

(1) Semen samples were collected from presumably fertile men, who were randomly recruited and had fathered at least one child, and suspected infertile men.

(2) The suspected subfertile male population was randomly recruited from the Urology Division and was defined as follows: three consecutive (biweekly or monthly) abnormal semen analyses that showed one or more abnormalities in the following three parameters: sperm count < 20 million/ml; motility at 2 h after collection $< 40\%$ and morphology $< 60\%$ normal forms. The female partner had normal ovulatory cycles, normal luteal functions, optimal cervical factors, patient fallopian tubes and absence of coital difficulties.

(3) Specimens were collected by masturbation and allowed to liquefy at room temperature for about 30 min.

(4) Semen analysis was done on Cell-Soft semen analyzer (Cryo Resources, New York, N.Y.) by taking a 5-µl specimen on a Makler counting chamber.

(5) Semen specimens were equally aliquoted into three polystyrene (15-ml) conical centrifuge tubes. Each aliquot was mixed in a 1:3 ratio with BWW medium supplemented with 0.3% BSA (Fraction V, Sigma) and pH adjusted to 7.4 before use.

(6) The tubes were centrifuged at $600\,g$ for 5 min in a tabletop centrifuge. Supernatant was aspirated by sterile pasteur pipette.

Table 1. Substances and proposed mechanisms of action

Substance	Proposed mechanism
Caffeine [1, 2]	increases intracellular cAMP increases ATP utilization
Pentoxifylline [3]	increases intracellular cAMP increases ATP utilization
Platelet-activating factor [4]	increases intracellular calcium increases intracellular ATP
Relaxin [5]	increases intracellular cAMP
Kallikrein/bradykinin [6]	acceleration of glucose and fructose across the sperm membrane and thus stimulation of sperm metabolism regulation of phosphofructokinase
Calcium [7]	increases intracellular calcium
Creatinine phosphate [7]	rephosphorylation of ADP to ATP in sperm tail
Prostaglandin E_2 [8]	increases intracellular cAMP

(7) Pellets were resuspended in BWW medium containing different concentrations of theophylline (0, 10, 22 mM). The final sperm concentration was adjusted to 10 million/ml.

(8) Sperm aliquots without theophylline served as controls.

(9) Spermatozoa suspensions were then incubated at 37 °C in horizontal position under the air atmosphere for 5 h.

(10) Eggs were collected from superovulated female Syrian golden hamsters by routine methods. Zona-free eggs were washed three times in fresh BWW medium before incubation with the treated spermatozoa.

(11) At the end of overnight incubation period, spermatozoa suspensions were diluted with equal volumes of BWW medium and centrifuged at 300 g for 3 min to remove the theophylline. The control tube received the same treatment.

(12) The supernatants were aspirated and the pellets were resuspended in fresh BWW medium to give a final concentration of 10 million/ml.

(13) Duplicates of 0.1-ml aliquots of the spermatozoa suspensions were placed in sterile plastic Petri dishes and were covered with mineral oil.

(14) Twenty to 25 zona-free ova were added to each droplet and the mixture was incubated in 5% CO_2 incubator at 37 °C.

(15) After 6 h of incubation, the ova were removed and washed in BWW medium twice before examination with a phase contrast microscope at 20 × magnification.

(16) Ova were recorded as penetrated when swollen human spermatozoal heads attached to tail were seen within the cytoplasm.

(17) Percent penetration was defined as the number of eggs penetrated/total number of eggs scores × 100.

Table 2. Effect of sperm washing with theophylline

Group	Theophylline concentration, mM	Mean percentage of penetration
Patients (n = 10)	0	16[1]
	10	46*
	20	51*
Donors (n = 9)	0	30
	10	50*
	20	52*

* p value <0.001 when compared to control, untreated (0 mM) samples.
[1] Egg penetration rate $<25\%$ was considered in infertile range in this study.

Results

Previous investigators had noted a beneficial effect of theophylline incubation on human sperm velocity [10]. Our study was designed to determine if theophylline incubation of human semen would result in enhanced sperm fertilizing capacity as measured by the hamster egg penetration test. The results of our study are summarized in table 2. In our laboratory a score of less than 25% of hamster eggs penetrated is considered to be in the infertile range. Incubation of semen with 10 mM of theophylline resulted in a statistically significant increase of percentage of eggs penetrated in both infertile and fertile patients, but was even more striking in the infertile group. Increasing the theophylline concentration to 20 mM did not result in further enhancement of egg penetration. However, the beneficial effect of theophylline washing was not seen in suspected infertile patients who had low ejaculatory volumes ($\leq$ 1 ml), a total motile sperm count of $\leq 10 \times 10^6$/ml or increased semen viscosity.

Discussion

Theophylline incubation of human semen appears to have a beneficial effect on the fertilizing capacity of the sperm of both fertile and suspected infertile men as measured by the hamster egg penetration test. No significant difference was seen between 10 and 20 mM theophylline concentrations. The mechanism of action of theophylline on washed semen samples is not well understood. Stimulation of intracellular levels of cAMP is a possible mechanism, but is at present unproven. Further investigation of sperm washing with theophylline and related agents offers the promise of improved sperm function.

References

1   Ruzich, J.V.; Gill, H.; Wein, A.J.; Van Arsdalen, K.; Hypolite, J.; Levin, R.M.: Objective assessment of the effect of caffeine and sperm motility and velocity. Fertil. Steril. *48:* 891–893 (1987).

2   Margalioth, E.J.; May, E.Y.; Navot, D.; Laufer, N.; Ovadia, J.; Schenker, J.G.: Effect of caffeine on human sperm penetration into zona-free hamster ova. Arch. Androl. *14:* 139–142 (1985).

3   Hellstrom, W.J.G.; Sikka, S.C.: Pentoxifylline stimulates the movement characteristics of cryopreserved semen. Surg. Forum *40:* 648–650 (1989).

4   Ricker, D.D.; Minhas, B.S.; Kumar, R.; Robertson, J.L.; Dodson, M.G.: The effects of platelet-activating factor on the motility of human spermatozoa. Fertil. Steril. *52:* 655–658 (1989).

5   Lessing, J.B.; Brenner, S.H.; Schoenfeld, C.; Goldsmith, L.T.; Amelar, R.D.; Dubin, L.; Weiss, G.: The effect of relaxin on the motility of sperm in freshly thawed human semen. Fertil. Steril. *44:* 406–409 (1985).

6   Scito, H.; Schill, W.B.: Temperature-dependent effects of the components of Gallibrein-Ginin system on sperm motility in vitro. Fertil. Steril. *47:* 684–688 (1987).

7   Fakih, H.; MacLusky, N.; DeCherney, A.; Wallimann, T.; Huszar, G.: Enhancement of human sperm motility and velocity in vitro: effects of calcium and creatine phosphate. Fertil. Steril. *46:* 938–944 (1986).

8   Colon, J.M.; Ginsburg, F.; Lessing, J.B.; Schoenfeld, C.; Goldsmith, L.T.; Amelar, R.D.; Dubin, L.; Weiss, G.: The effect of relaxin and prostaglandin E_2 on the motility of human spermatozoa. Fertil. Steril. *46:* 1133–1139 (1986).

9   Aitken, R.J.; Mattei, A.; Irvine, S.: Paradoxical stimulation of human sperm motility by 2-deoxyadenosine. J. Reprod. Fertil. *78:* 515–527 (1986).

10 Jiang, C.S.; Kilfeather, S.A.; Pearson, R.M.: The stimulatory effects of caffeine, theophylline, lipine-theophylline and 3-isobutyl-1-methylxanthine on human sperm motility. Br. J. Clin. Pharmacol. *18:* 258–262 (1984).

Dr. Ashok Agarwal, Division of Urology, Department of Surgery,
Brigham and Women's Hospital, 45 Francis Street, Boston, MA 02115 (USA)

Colpi GM, Pozza D (eds): Diagnosing Male Infertility.
Prog Reprod Biol Med. Basel, Karger, 1992, vol 15, pp 66–70

Comparison of Low Hypo-Osmotic Swelling Results to Fertilization Rates in an IVF-ET Program

Jerome H. Check[a], Aniela Bollendorf[b], Mike Lee[a], Beth Vetter[b], Mark Syrkin[b], Jeffrey S. Chase[b]

[a] University of Medicine and Dentistry of New Jersey,
Robert Wood Johnson Medical School at Camden,
Cooper Hospital/University Medical Center;
[b] Department Ob/Gyn, Division of Reproductive Endocrinology and Infertility,
Camden, N.J., USA

Previous studies have shown that in vivo pregnancies correlated with hypo-osmotic swelling (HOS) scores. Men who had subnormal motile densities (MD) but normal HOS scores achieved a high pregnancy rate with female factors corrected. No significant differences were seen in men with normal HOS scores $\geq 60\%$ and grey-zone HOS scores between 50 and 59%. However, there were no pregnancies in the female counterpart when the males HOS score was $< 50\%$ [1–3].

The data presented herein evaluates whether these patients with otherwise normal semen parameters but low HOS scores ($< 50\%$) have reduced fertilization in vitro and subsequent pregnancy rates.

Materials and Methods

Paired semen samples were analyzed at a 2- to 4-week interval prior to the in vitro fertilization (IVF) embryo transfer (ET) cycle. Sperm count, motility, morphology, viability and HOS were performed on both semen samples.

Normals for IVF semen analysis (SA) are: count $\geq 20 \times 10^6$/ml, motility $\geq 30\%$ and morphology $> 60\%$ normal forms [4]. The SA was manually assessed using a phase contrast microscope. The Makler chamber was used for count and motility results. Two stained semen smears were analyzed for sperm morphology. Sperm morphology was evaluated by one of two methods. Earlier the criteria suggested by the World Health Organization (WHO) was used [5]. This criteria was used for approximately one third of the patients. The remaining were evaluated using a more strict criteria report by Kruger et al. [6]. Strict morphology results were considered abnormal if there were $< 5\%$ normal forms.

Sperm viability was assessed using equal parts 0.5% Eosin Y stain and semen. The semen was then placed on a glass slide coverslipped and allowed to sit for 5 min at room temperature. The unstained sperm (viable) were counted microscopically. WHO suggests $> 50\%$ viable as normal [5]. The HOS was performed as previously described [7].

A total of 267 couples with at least one IVF-ET cycle were divided into 4 groups: group 1, normal HOS and normal SA; group 2, normal HOS and abnormal SA; group 3, subnormal HOS and normal SA; group 4, subnormal HOS and abnormal SA.

Follicular Stimulation and Embryo Retrieval
Follicular stimulation was accomplished by first suppressing endogenous gonadotropins with administration of leuprolide acetate (LA), 1 mg s.c. 7 days following ovulation, for a period of 10 days. Human menopausal gonadotropins (hMG) was administered intramuscularly at a dosage of 150 IU twice daily, starting on day 11 of the cycle for 3 days. The hMG was then reduced to 225 IU/day for 1 day, then 150 IU daily thereafter. Patients were monitored daily via vaginal ultrasonography and serum estradiol (E_2) measurement. At the point when a serum E_2 level of at least 600 pg/ml and when at least 3 follicles over 16 mm diameter were observed, hCG (human chorionic gonadotropins) was administered at a dosage of 10,000 IU i.m. Oocyte retrieval was performed via vaginal ultrasound 30–32 h post-hCG.

The retrieved oocytes were classified into four categories: immature, mature, postmature, or atretic. All preovulatory oocytes were incubated in Ham's F-10 medium (Gibco, Grand Island, N.Y.), supplemented with 5% bovine serum albumin (BSA) (Sigma Chemical Co., St. Louis, Mo.), pH 7.4 at 275–280 mosm/kg H_2O. Motile sperm concentrations of $5–10 \times 10^4$/ml were used for inseminations. Higher sperm concentrations of $1–2.5 \times 10^5$/ml were used for oligospermic patients. Semen was centrifuged at 400 *g* for 8 min in a $2\times$ volume of F-10 + BSA and the resulting pellet overlaid with 0.2–0.5 ml modified F-10. Motile sperm were allowed to swim up into the overlying medium for 30–60 min. Following removal of the motile fraction, semen parameters were measured and used for insemination. One to 5 embryos were transferred on the second day after retrieval. When there were more than 5 embryos, the remaining were cryopreserved for subsequent transfer, using a modification of the one-step method. All patients who underwent ET received supplemental progesterone (25 mg in oil i.m.) daily, starting on the day of retrieval and continuing until a negative β-hCG test result was obtained.

Sperm count and motility were evaluated on the day of IVF retrieval. This count and motility was used along with the initial morphology result to place the IVF cycle into its appropriate group.

Table 1. Comparison of semen results of the four patient groups

	Group 1	Group 2	Group 3	Group 4
Patients	221	44	10	5
Mean count, $\times 10^6$/ml	98 ± 98.2	64.2 ± 62.2	112.6 ± 62.13	39.77 ± 33.29
Mean motility, %	64.5 ± 13.35	53.1 ± 21.85	58.3 ± 12.58	53.18 ± 25.99
Mean WHO criteria morphology % normal forms	65.5 ± 11.7	43.4 ± 15.3	69.5 ± 1.5	35.0 ± 4.1
Mean strict morphology % normal forms	13.9 ± 16.1	4.9 ± 5.6	14.3 ± 4.2	02.5 ± 0.5
Mean HOS, %	74.2 ± 8.9	70.0 ± 9.8	41.2 ± 5.9	40.9 ± 16.6

Table 2. Comparison of IVF results of the four patient groups

	Group 1	Group 2	Group 3	Group 4
Patients	221	44	10	5
Retrievals	377	60	13	11
Mean number of eggs inseminated	9.05 ± 6.4	8.46 ± 5.56	9.3 ± 3.72	8.6 ± 5.3
Mean number of embryos transferred	3.26 ± 4.76	2.08 ± 2.19	2.92 ± 1.6	2.09 ± 2.38
Mean number of embryos frozen	1.32 ± 2.57	0.76 ± 1.7	1.38 ± 2.2	1.18 ± 1.78
Pregnancies	59	2	2	0
Pregnancy rate/retrieval, %	15.7	3.3	15.4	0
Abortion rate, %	19	100	50	
Mean fertilization rate, %	52.65 ± 31.37	33.3 ± 34.0	36.3 ± 31.8	33.63 ± 42.9

Results

In the control group (group 1) the count, motility and percent swelling (HOS) are higher than in groups 2, 3 and 4 (table 1).

In table 2 the number of eggs inseminated, number of embryos transferred and number of embryos frozen are essentially equal in all groups. Also, in groups 2, 3 and 4 the percent fertilization in vitro was essentially the same. Percent fertilization in group 1 (control group) was twice as high as in groups 2, 3 and 4. The pregnancy rate in groups 1 and 3 per retrieval is also much greater than that in groups 2 and 4. Differences between groups 1 and 3 are due to differences in HOS parameters only, i.e. fertilization rate. But at the same time, pregnancy rate per retrieval does not follow this trend.

Discussion

It has been reported by various IVF centers that an abnormal semen analysis is associated with low fertilization rates in vitro [4]. In our study presented herein, groups 1 and 3, both having a normal semen analysis, should have similar results. Interestingly, group 3 (normal SA but subnormal HOS scores) had lower fertilization rate than group 1. Group 3 also had similar fertilization rate as groups 2 and 4 (both containing men with subnormal semen analysis).

These data suggest that a subnormal HOS score does suggest reduced fertilization rates in vitro but 0% pregnancy rate in vivo. It appears to be an important semen parameter that can indicate a male factor contribution to the low fertilization rates in vitro which could have been classified as unexplained infertility.

Studies are currently being performed to evaluate a larger group of patients with subnormal HOS scores in vitro. These data suggest that in vitro fertilization may be an alternate approach to the treatment of men with subnormal HOS scores.

References

1   Check, J.H.; Epstein, R.; Nowroozi, K.; Shanis, B.S.; Wu, C.H.; Bollendorf, A.: The hypo-osmotic swelling test as a useful adjunct to the semen analysis to predict fertility potential. Fertil. Steril. *52:* 159–161 (1989).
2   Check, J.H.; Nowroozi, K.; Wu, C.H.; Bollendorf, A.: Correlation of the semen analysis and the hypo-osmotic swelling test with subsequent pregnancies: A preliminary report. Arch. Androl. *20:* 256–260 (1988).
3   Check, J.H.; Nowroozi, K.; Wu, C.H.; Bollendorf, A.: The Clinical Significance of the Hypo-Osmotic Swelling (HOS) Test. Meeting of the Pacific Coast Fertility Society, Palm Springs, Calif., April 1989.
4   Yates, C.A.; De Kretser, D.M.: Male factor infertility and in vitro fertilization. J. In Vitro Fertil. Embryo Transfer *4:* 141–147 (1987).

5   World Health Organization Laboratory Manual for the Examination of Human Semen and Semen-Cervical Mucus Interaction (Cambridge University Press, Cambridge 1987).

6   Kruger, T.F.; Acosta, A.A.; Simmons, K.F.; Swanson, R.J.; Matta, J.F.; Oehninger, S.: Predictive value of abnormal sperm morphology in in vitro fertilization. Fertil. Steril. *49:* 112–117 (1988).

7   Jeyendran, R.S.; Van der Ven, H.H.; Perez-Palaez, M.; Crabo, B.G.; Zaneveld, L.J.D.: Development of an assay to assess the functional integrity of the human sperm membrane and its relationship to other semen characteristics. J. Reprod. Fertil. *70:* 219–228 (1984).

Jerome H. Check, MD, 7447 Old York Road, Melrose Park, PA 19126 (USA)

Colpi GM, Pozza D (eds): Diagnosing Male Infertility.
Prog Reprod Biol Med. Basel, Karger, 1992, vol 15, pp 71–74

Evaluation of the Kruger Strict Method for Sperm Morphology in Predicting Infertile Males

Jerome H. Check[a], Marie Press[b], Aniela Bollendorf[b], Todd Blue[b]

[a] The UMDNJ, Robert Wood Johnson Medical School at Camden,
Cooper Hospital/University Medical Center;
[b] Department Ob/Gyn, Division of Reproductive Endocrinology and Infertility,
Camden, N.J., USA

Some researchers consider sperm morphology as the best parameter of the semen analysis in predicting fertilization potential [1]. Sperm morphology (as determined by the World Health Organization) has been a semen variable used to evaluate the male factor with the outcome of in vitro fertilization of human eggs.

Recently, a new strict method used by Kruger et al. [2] has been demonstrated to correlate with success of ova fertilization by in vitro methods. They found a very poor fertilization rate and subsequent pregnancy rate when the strict method demonstrated only 4% or less of the sperm to have normal morphology. However, by increasing the number of sperm inseminated per egg in men with poor morphology, improved pregnancy rates have been recorded [3]. They demonstrated a very low fertilization rate in men with poor Kruger tests when the standard 50,000 motile sperm/oocyte was added but increased the percent of oocyte fertilization significantly to 62.6% by increasing the number of sperm inseminated to 500,000 motile sperm/oocyte. Interestingly, however, poor pregnancy outcome still occurred.

A study was conducted to determine if in a group of infertile couples not undergoing in vitro fertilization where a female factor has been identified and seemingly corrected, would a correlation with normal morphology in the male partner be seen in the couples achieving pregnancies and would poor morphology correlate with those failing to conceive? Furthermore, the Kruger strict criteria would be compared to WHO morphologic standards to see if the former better identifies the subfertile males.

Table 1. Kruger method criteria for sperm morphology

Normal	Must have smooth oval shape Well-defined acrosome, consisting of 40–70% of the head Cytoplasmic droplets less than half the size of the sperm head Length of head 5–6 μm Diameter of head 2.5–3.5 μm No neck, midpiece or tail defects
Slight amorphous	Head diameter 2.0–2.5 μm Slight abnormalities in the shape of the head Must have a normal acrosome Neck – debris of thickness, but normal head
Severe amorphous	No acrosome or <30% or >70% Completely abnormal shape Neck – bend in neck or midpiece

Materials and Methods

Two separate studies were conducted. The first study consisted of 41 infertile couples (group 1) where the male counterpart's semen analysis was evaluated for morphology by the Kruger strict criteria (table 1). The second study consisted of 145 couples (group 2) where the WHO criteria were used for assessing morphology. The two groups were composed of entirely different patients.

A requirement for both groups was that a female factor be identified and seemingly corrected. Motile densities were determined using a Makler chamber. Men with less than 5×10^6 sperm/ml were excluded from these studies. The pregnancy rates in 6 months were determined.

Results

The correlation of strict morphologic changes in group 1 males and 6-month pregnancy rates in their female partners are seen in table 2. Though the numbers are small, just as many if not more patients with 5–14% normal forms achieved pregnancies in their wives compared to those wives of men with 14%. There were not enough cases yet to fully investigate the less than 5% category. Though not part of this study, so far 2 men out of 233 have achieved a pregnancy in their wives with strict morphological criteria under 5%. Of interest was also the fact that the motile density did not effect the pregnancy rate. A total of 24/33 men

Table 2. Correlation of morphology (as determined by Kruger's stricter criteria) and pregnancy rates (6 months)

	MD $> 10 \times 10^6$/ml			MD $< 10 \times 10^6$/ml		
	Kr > 14	Kr 5–14	Ke < 5	Kr > 14	Kr 5–14	Ke < 5
Patients	24	8	1	2	6	0
Pregnancies	9	4	0	1	4	0
Pregnant, %	33	50	0	50	67	0

Table 3. Correlation of morphology (as determined by WHO criteria) and pregnancy rates (8 months)

	MD $> 10 \times 10^6$/ml		MD $< 10 \times 10^6$/ml	
	normal WHO	subnormal WHO	normal WHO	subnormal WHO
Patients	98	22	20	8
Pregnancies	83	18	20	6
Pregnant, %	87	82	100	75

(73%) were found normal. Though 84.6% conceived when the motile density by WHO was normal; nevertheless so did 82% with abnormal WHO morphology as seen in table 3. Again, motile densities failed to correlate with achievement of pregnancies in the female partners.

Discussion

With in vitro fertilization (IVF), theoretically, fertilization may be achieved by a lower percentage of normal sperm since the sperm are placed almost immediately together with the ova directly, rather than traverse the large distance reproductive tract and be required to survive many hours before the egg is naturally prepared and in position for fertilization. Though Kruger has found only 4% or lower correlates with IVF fertilization, the possibility exists that this number is low for situations more demanding for the sperm, i.e. in vivo fertilization. However, there were no statistically significant differences (p = 0.41) between men > 14% normal versus 5–14% concerning their fertilization potential.

Whether we will corroborate the conclusions from IVF data in the in vivo system, i.e. that morphology < 5% can distinguish fertile from subfertile males, will have to await a larger series. The reason for the higher pregnancy rates in female partners of group 2 men versus group 1 is not clear but may be related to a fortuitous difference in average age for the females (36.5 in group 1 and 30.3 in group 2). The data was surprising in that even males with the combination of subnormal motile densities (MD) with strict criteria only 5–14% did not demonstrate a reduced fertility rate in comparison to normal MD and normal morphology.

References

1   Rogers, B.J.; Bentwood, B.J.; Van Campen, H.; Helmbrecht, G.; Soderdahl, D.; Hale, R.W.: Sperm morphology assessment as an indicator of human fertilizing capacity. J. Androl. *4:* 119–125 (1983).

2   Kruger, T.F.; Acosta, A.A.; Simmons, K.F.; Swanson, R.J.; Matta, J.F.; Oehninger, S.: Predictive value of abnormal sperm morphology in in vitro fertilization. Fertil. Steril. *49:* 112–117 (1988).

3   Oehninger, S.; Acosta, A.A.; Morshedi, M.; Veeck, L.; Swanson, R.J.; Simmons, K.; Rosenwaks, Z.: Corrective measures and pregnancy outcome in in vitro fertilization in patients with severe sperm morphology/abnormalities. Fertil. Steril. *50:* 283–287 (1988).

Jerome H. Check, MD, 7447 Old York Road, Melrose Park, PA 19126 (USA)

Colpi GM, Pozza D (eds): Diagnosing Male Infertility.
Prog Reprod Biol Med. Basel, Karger, 1992, vol 15, pp 75–85

Endpoint of First Stage of Zona pellucida-Induced Acrosome Reaction in Mouse Spermatozoa Characterized by Acrosomal H⁺ and Ca²⁺ Permeability: A Possible Model for the Human System

Michael A. Lee[a], *Bayard T. Storey*[b], *Jerome H. Check*[a]

[a] Department of Obstetrics and Gynecology UMNDJ/Robert Wood Johnson
Medical School, Camden, N.J.;
[b] Departments of Obstetrics and Gynecology and Physiology,
University of Pennsylvania, School of Medicine, Philadelphia, Pa., USA

For the capacitated mouse spermatozoon to fertilize the egg, the zona-induced acrosome reaction (ZIR) must occur and the spermatozoon must pass into the perivitelline space. The time course of the ZIR has been resolved into two stages [Lee and Storey, 1985] by use of three fluorescent probes: chlortetracycline (CTC) [Saling and Storey, 1979; Ward and Storey, 1984], 9-amino-3-chloro-7-methoxyacridine (ACMA), and 9-(N-dodecyl) aminoacridine (NDAA). The CTC assay shows three sequential fluorescent patterns on the sperm head; B-to-S and S-to-R. The B pattern has been shown to be indicative of capacitated, acrosome-intact spermatozoa. The CTC S pattern, an intermediate pattern, is that of a cell which has lost the intra – to extracellular H⁺ gradient as shown by the aminoacridine pH probes. The CTC R is that of an acrosome-reacted spermatozoon. The B-to-S transition occurs concurrently with the loss of the sperm head H⁺ gradient. In this study we posed the question: Is the loss of the H⁺ gradient accompanied by an influx of calcium into an intracellular compartment of lower Ca²⁺ or does the influx of Ca²⁺ occur prior to the loss of the H⁺ gradient? The fluorescent probe fura-2 was used to measure the cell population kinetics of Ca²⁺ influx into the sperm head by loss of fura-2, head fluorescence. Nigericin, a K^+/H^+ exchange ionophore, was used to discharge the H^+ gradient prior to the B-to-S transition. We present evidence in this poster that the key event is Ca^{2+} influx, with loss of H^+ gradient secondary to this event.

In the mouse system it has been demonstrated that a guanine nucleotide-binding regulatory protein (Gi) is involved in the ZIR [Endo et al., 1988] by block of the ZIR by pertussis toxin (IAP). In the human system we have approached the system of sperm-egg interaction by use of solubilized zona proteins. IAP blocks the ZIR in human and has no effect on the spontaneous or ionophore-induced aerosome reaction. This implicates a Gi. These Gi proteins occupy critical roles as signal-transducing elements in coupling many ligand-receptor interactions with the generation of intracellular second messengers. These heterotrimeric plasma membrane-associated proteins are composed of distinct α-subunits, which contain a GTP-binding domain, and more highly conserved β- and γ-subunits. The α-subunits can be covalently modified by ADP-ribosylation in the presence NAD^+ by the action of a variety of bacterial toxins, including cholera toxin, pertussis toxin, and botulinum toxim. Sperm suspensions were preincubated with IAP and assayed for acrosomal status. The ability to undergo toxin-catalyzed ADP-ribosylation has been used by many investigators to identify general classes of G proteins. The ability to undergo toxin-catalyzed ADP-ribosylation has been used by many investigators to identify general classes of G proteins. We present preliminary evidence to suggest the a Gi protein is operational in the ZIR in human spermatozoa.

Materials and Methods

Sperm Capacitation and Binding to Structurally Intact Zonae pellucidae

Sperm suspensions were prepared as described by Lee and Storey [1985]. Mechanically isolated, structurally intact zonae pellucidae were prepared by forcing cumulus-free eggs through a narrow bore micropipette (i.d. = 0.75 diameter of egg with zonae). All incubations were carried out in HMB containing 4 mg/ml bovine serum albumin (BSA) at 37 °C in a humidified atmosphere of 5% CO_2 in air. Sperm were capacitated at a concentration of 10–25×10^6 cells/ml for 90 min in the above medium containing 4 mM fura-2, AM or no fura-2, AM for NDAA experiments. After capacitation, the final sperm concentration was adjusted to 2×10^5 cells/ml by adding an aliquot of the labeled sperm suspension to a 100-μl drop containing 100–150 mechanically isolated zonae pellucidae. The point of addition of sperm to the isolated zonae was taken to be the zero time point. Sperm were allowed to bind to the zonae for 15 min, at which time

the zonae with bound sperm were washed 3 times in HMB using a wide-bore micropi-
pette to remove loosely associated sperm. For human experiments the culture medium
HTF and modified HTF used in the handling of oocytes was purchased from Irvine
Scientific. The culture medium used for incubation and washing of spermatozoa was
Ham's F-10 purchased from Gibco Laboratories. The medium contained BSA at 5 mg/ml
in all experiments.

Human sperm ejaculates were obtained from healthy, known fertile, adult donors
via masturbation after 36–48 h abstinence. The ejaculates were allowed to liquify at
room temperature for 15–30 min. All ejaculates had initial parameters of at least 65%
motility, 60% motility, 60% normal morphology, and cell counts of 35 $\times$ 10^6/ml or
more. After the initial ejaculate had liquified, it was mixed with 3 volumes F-10 containing 5 mg/ml BSA. The sperm was washed by centrifugation in 12-ml conical tubes at 600
g for 10 min. The supernatant was removed and the sperm pellet gently overlaid with 2
ml Ham's F-10 containing 5 mg/ml BSA. The centrifuged tube was inclined at 45° and
the motile sperm allowed to swim up out of the pellet for 30 min. This technique is a
modification of the method of Overstreet et al. [1976]. The sperm sample obtained by
this method showed motility percentages over 90% consistently. In experiments, IAP
was added to the sperm suspension to a final concentration of 1 mg/ml. Sperm suspensions were incubated for 3 h to capacitate prior to addition of A23187 or solubilized
human zona proteins.

Preparation of Mechanically Isolated, Structurally Intact Human Zonae pellucidae

Oocytes were obtained from IVF patients whom had at least one oocyte fertilized
per treatment cycle to exclude possible zona abnormalities. All zona were removed from
oocytes which were previously inseminated and did not fertilize within 72 h postretrieval.
Zona were removed from oocytes judged as mature, immature, or postmature at the time
of oocyte retrieval. All oocytes were cultured in HTF medium supplemented with 5 mg/ml
ultrapure BSA. Each oocyte was removed from the culture dish using a drawn micropipet
and placed into 100 μl of Hepes-buffered (10 mM) HTF, (MHTF) at pH 7.4 in a lux
50-well culture dish. The oocyte was transferred through three 100-μl drops to remove
residual cumulus and serum albumin. The oocyte was transferred to a fourth drop of
MHTF, pH 7.4, containing 1 mg/ml EDTA and 10 mg/ml Lima bean trypsin inhibitor.
The oocyte was transferred to a fourth drop of the above medium and aspirated into a
pipet 0.75 diameters of the egg. This facilitates rupture of the cytoplasm and extrusion of
the intact zona pellucida. The intact zona is then transferred through four washes of the
above medium. The intact zona is then placed into 100 μl of 1% Triton X-100 buffered
with 10 mM ADA, 1 M NaCl and 1 mM benzamidine in DH_2O for 1 min. The intact zona
are then transferred through four washes of 1 mM benzamidine in DH_2O and twice
through DH_2O. The zona was then placed in a 1-ml nunc vial containing 50 μl of glycerol
supplemented with 20 mg/ml ultrapure BSA and stored at 0 °C until used. Induction of
the acrosome reaction was initiated after a 3-hour incubation period in medium containing 5 mg/ml ultrapure BSA. Two methods were used to induce acrosome reactions: treatment with A23187 [Wolf et al., 1985; Byrd and Wolf, 1986], or incubation with acid-
solubilized human zona pellucida. Human zona were removed from oocytes that were not
fertilized in vitro and stored as described previously. 30–50 zonae were polled into one
tube and washed by centrifugation 3$\times$ with culture medium. The supernatant was
removed and acidified (pH 4.0) Ham's F-10 without BSA was added to bring the zonae

concentration to 10 zonae/µl. The zonae were incubated at 37 °C for a period of 1 h in order to dissociate the zona proteins. The final solution at pH 7.4 containing solubilized zonae pellucidae was added immediately to an appropriate volume of capacitated sperm suspension to achieve a final concentration of 2 zonae/µl [Cross et al., 1986]. In the treatment with A23187, stock A23187 was added to an appropriate volume of capacitated spermatozoa to achieve a final concentration of 10 μM which equals 0.1% DMF (v/v). DMF alone was added to a second aliquot to 0.1% (v/v). Incubation was continued at 37 °C for an additional 2 h after the addition of either A23187 or zonae proteins. At the end of the incubation with either ionophore or solubilized zonae (5 h total), 5-µl aliquots were removed from each sample for the CTC assay. The remainder of the samples were prepared for the FITC-PSA lectin assay.

Fluorescence Assays

Fura-2, AM was dissolved as a 10-mM stock and aliquoted into 10-ml vials in DMF and kept in a light-shielded container at 4 °C at all times. NDAA was dissolved at 10 mM in DMF and kept under similar conditions. In a third set of experiments the ionophore nigericin was added to the final wash drop containing zonae with bound labeled sperm (fura-2,AM or NDAA) to a concentration of 2.5 mg/ml. The binding of sperm to the isolated zonae pellucidae was scored by phase-contrast microscopy ($\times$ 400), with a Nikon Optiphot microscope equipped with interchangeable phase-contrast and epifluorescence optics. The bound sperm were then examined by epifluorescence microscopy ($\times$ 400) for fura-2,AM or NDAA fluorescence. Mn^{2+} was added to final concentration of 0.25 mM in fura-2,AM experiments. NDAA solution was added to a final concentration of 10 mM in those experiments. Control zonae had an equal volume of DMF (NDAA carrier) added. In no experiment did the DMF concentration exceed 0.1%. After 5 min incubation, 10 zonae with bound sperm (fura-2,AM or NDAA labeled) were placed on a slide and covered with a coverslip. The remaining zonae with bound sperm were placed in the incubator for later observations. At the time of assay, 10 washed zonae with bound labeled sperm were placed on a slide immediately after which a coverslip was attached. These slides were placed on a warmed (37 °C) stage and viewed at 30-min intervals with phase-contrast and epifluorescence optics. Fluorescence loss was observed at the time intervals postbinding. At the time of assay, 9.9. µl of HMB containing 10 washed isolated zonae with bound fura-2,AM labeled sperm or NDAA were added to a warmed slide. Fluorescence excitation was at 405 nm with a half band width of 30 nm, using the excitation unit of the spectrum analyzer and the Optiphot 'V' filter cassette with excitation transmission between 395 and 425 nm with peak at 405 nm for NDAA and fura-2,AM. The emission was filtered through the DM 445 dichroic mirror of the filter cassette and a 470K guard filter. The 405 nm excitation of the fura-2,AM corresponds to the Ca^{2+} free form of the probe [Grynkiewicz et al., 1985]. Controls for fura-2,AM experiments contained an equal amount of DMF (fura-2,AM carrier).

The CTC stock solution was prepared as previously described [Ward and Storey, 1984; Lee and Storey, 1985, 1989]. At the appropriate times during incubations (see Discussion), 5 µl of sperm suspension was placed on a warmed (37 °C) slide and an equal volume of CTC stock solution added, followed within 30 s by 0.05 µl 12.5% gluteraldehyde in 1 M Tris buffer (pH 7.8). The slides were kept in a humidified, light-shielded container until scoring. The sperm were examined at 400 $\times$ for CTC fluorescence patterns with an Olympus BHZ fluorescent microscope equipped with phase-contrast and epifluorescent optics. A minimum of 200 spermatozoa were scored for each determination for each time point of individual experiments.

The fluorescent FITC-PSA lectin was prepared as previously described by Cross et al. [1986]. At the appropriate times postincubation the spermatozoa were centrifuged at 600 g for 10 min and the supernatant removed. 50–200 µl of methanol was added to the pellet and resuspended. The methanol spermatozoa suspension was allowed to sit at 4 °C overnight. 20-ml aliquots were placed onto cleaned slide and allowed to air dry. 20 ml of FITC-PSA lectin at a concentration of 100 mg/ml was added to each slide and allowed to incubate in a humidified, light-shielded container for 10 min. Unbound FITC-PSA was rinsed away by dipping each slide into a beaker of deionized water for approximately 15 s. Water was drained from the slide and the remainder wiped away from the center where the labeled sperm are located. A drop of mounting medium is added to the slide and a cover glass attached. A minimum of 200 cells were scored for each determination of the percentage of FITC-PSA lectin binding spermatozoa in each time point for each experiment.

Results

The Ca^{2+} probe fura-2 gives fluorescence over the entire mouse sperm cell after preincubation with fura-2 to 'load' the dye into the cell. But a more regionalized fluorescence over the anterior portion of the sperm head is seen in the presence of 0.25 mM Mn^{2+} to act as quencher for extracellular fura-2. The fluorescence over the anterior head represents fura-2 in the Ca^{2+} free form, trapped in an intarcellular space inaccessible to divalent ions, e.g. the fluorescence quenching divalent cation Mn^{2+} [Grynkiewicz et al., 1985]. This is evidently the acrosome itself or acrosome plus periacrosomal space. The head fluorescence is lost after time in sperm bound to isolated, structurally intact mouse zonae pellucidae (fig. 1). In the absence of added Mn^{2+}, fluorescence over the entire cell remains unchanged with time under these conditions. The kinetics of the loss of fura-2 fluorescence from the acrosomal region gives a characteristic 30-min lag period previously demonstrated with the fluorescent probes CTC, ACMA, and NDAA (fig. 2). The time course kinetics of loss of fura-2 fluorescence is not disturbed by the addition of the ionophore nigericin, which abolishes the head fluorescence of the pH probes ACMA and NDAA (fig. 3), in zona bound mouse permatozoa. The ionophore nigericin acts by equilibrating the H^+ gradient across membranes by K^+/H^+ exchange. Loss of ACMA and NDAA fluorescence indicates loss of this gradient between the intracellular space over the anterior head and the suspending medium [Lee and Storey, 1985, 1989]. Fura-2 enables us to separate the loss of the H^+ gradient from

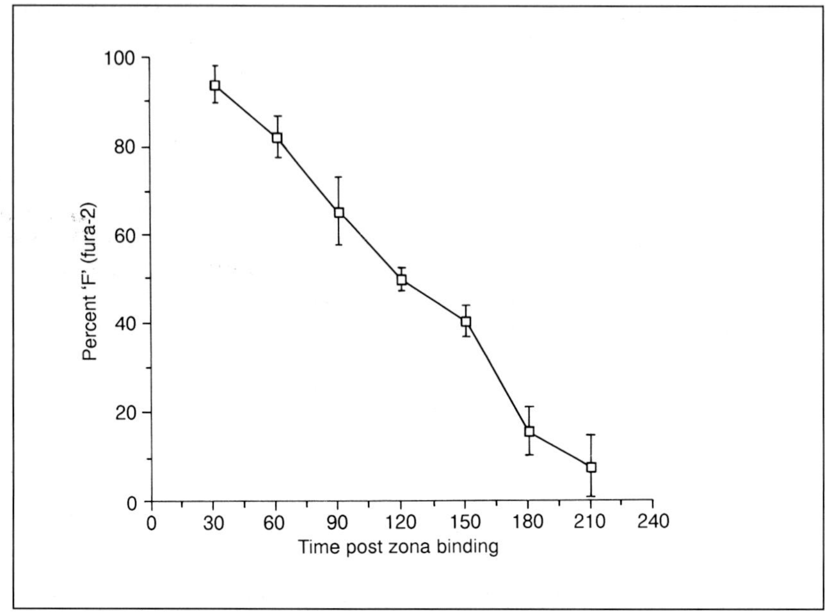

100
80
60
40
20
0
Percent 'F' (fura-2)
0 30 60 90 120 150 180 210 240
Time post zona binding
1

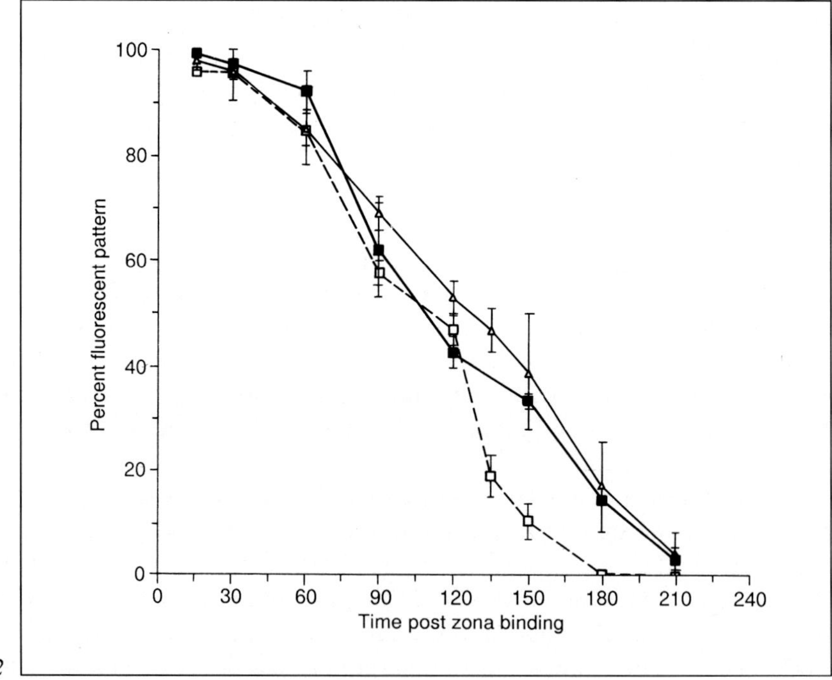

100
80
60
40
20
0
Percent fluorescent pattern
0 30 60 90 120 150 180 210 240
Time post zona binding
2

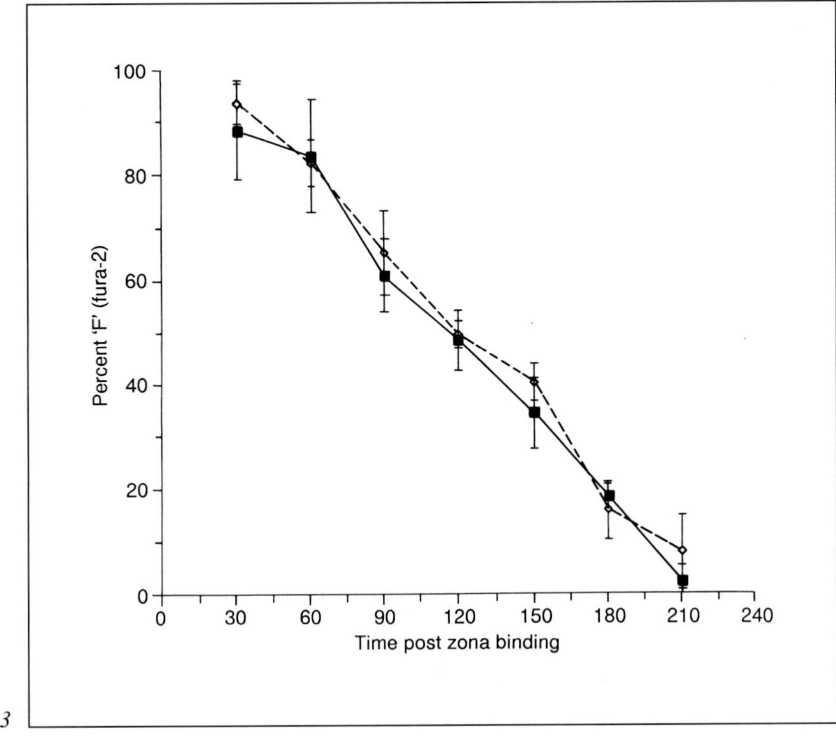

Fig. 1. Time course of the percentage of capacitated mouse sperm bound to isolated zona pellucida showing pattern 'F' (fura-2) head fluorescence. Note the characteristic 30-min 'lag' time prior to the onset of loss of fura-2 fluorescence loss. Each point represents the mean ± SD of 5 replicate experiments with a minimum of 10 zona pellucida scored per experiment per point.

Fig. 2. Time course of the percentage of capacitated mouse sperm bound to isolated zona pellucida showing pattern 'B' (CTC), 'A' (ACMA), or 'N' (NDAA). Open squares denote scoring of sperm bound using the CTC fluorescence pattern 'B'. Closed squares denote scoring of sperm bound using the ACMA fluorescent assay having the fluorescent pattern 'A'. Open triangles denote scoring of bound sperm using the NDAA fluorescent assay having the fluorescent pattern 'N'. Each point represents the mean ± SEM of 20 replicate experiments, with a minimum of 10 zona pellucida scored per experiment per point.

Fig. 3. Time course of the percentage of capacitated mouse sperm bound to isolated zona pellucida showing pattern 'F' (fura-2) head fluorescence ± nigericin. Open diamonds denote control spermatozoa loaded with fura-2. Closed boxes represent sperm scored in the presence of the ionophore nigericin. Note the characteristic 30-min 'lag' time prior to the start of loss of fura-2 fluorescence. Each point represents the mean ± SD of 5 replicate experiments, with a minimum of 10 zona pellucida scored per experiment per point.

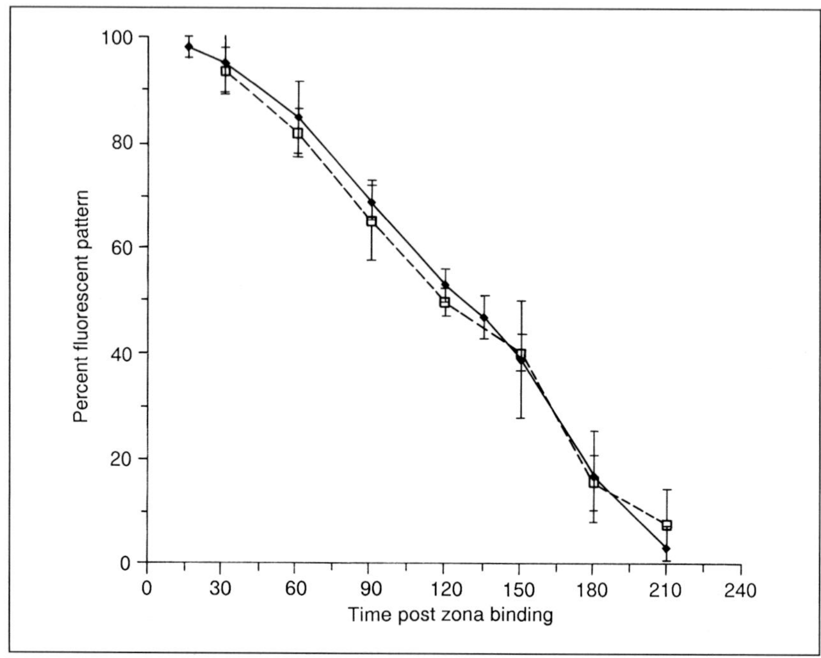

Fig. 4. Time course of the percentage of capacitated mouse sperm bound to isolated zona pellucida showing pattern 'F' (fura-2), and pattern 'N' head fluorescence. Open boxes denote spermatozoa scored after loading with the Ca^{2+} probe fura-2. Closed diamonds denote spermatozoa scored using the fluorescent pH probe NDAA. Each point represents the mean ± SD of 5 replicate experiments with a minimum of 10 zona pellucida scored per experiment.

the influx of Ca^{2+} in the early stages of the ZIR in mouse spermatozoa. The implication is that loss of the H^+ gradient as it occurs during the ZIR is a consequence of Ca^{2+} influx, not a precursor to it (fig. 4).

In the human system we have used solubilized zona proteins to demonstrate the ZIR. This reaction is inhibited by preincubation during capacitation of the spermatozoa with IAP which ADP ribosylates the α-subunit of the Gi protein heterodimer. The spontaneous and ionophore-induced acrosome reactions are not affected by IAP (fig. 5). The inhibition by IAP of the ZIR implicates a Gi protein is involved in the modulating of this reaction. The loss of acrosomal status was monitored using the CTC and PSA-FITC fluorescent labeling probes. Both fluorescent probes give similar results in monitoring human spermatozoa acrosomal status.

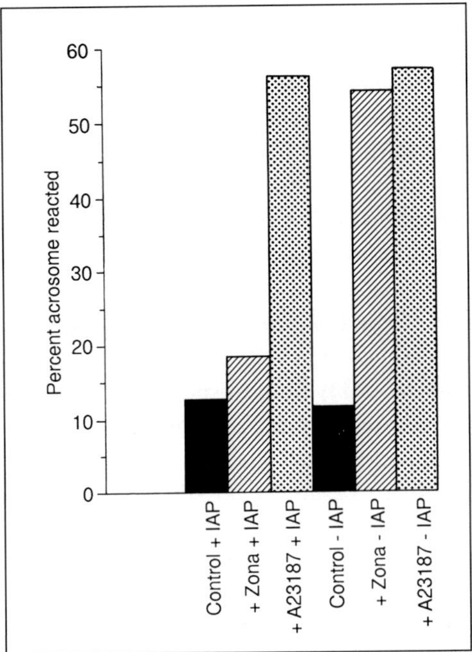

Fig. 5. Percentage of human spermatozoa acrosome reacted scored via FITC-PSA lectin ± 1 mg/ml IAP + 2 zona/ml or 10 mM A23187. Dark bars represent control sperm incubated 3 h ± IAP and incubated an additional 2 h. Striped bars represent sperm incubated 3 h ± IAP 2 zona/ml for an additional 2 h. Note the lack of acrosome reaction in the IAP-treated sperm. Mottled bars represent sperm preincubated 3 h ± IAP + 10 mM A23187 for an additional 2 h.

Discussion

In this study we have addressed the question of intracellular Ca^{2+} localization in the mouse sperm by observation of single cells bound to zonae. Localization was achieved by utilizing the fluorescent quenching property of Mn^{2+} to eliminate extracellular fluorescence. The fluorescence observed in the presence of Mn^{2+} must be from fura-2 in an intracellular compartment not accessible to extracellular Mn^{2+} and, by extension, not accessible to extracellular Ca^{2+}. From its location on the sperm head, we infer that the fluorescence is emanating from the acrosome and or acrosomal space overlying the acrosome. This fluorescence has the same regional location as that of monoclonal antibodies.

It should be noted that the 4 μM fura-2,AM used during the loading procedure was immediately diluted to 0.08 μM upon adding the sperm to the drops of media containing the zona and this dilution was followed by three washes with fresh media to insure reduction fo free fura-2,AM.

The onset of membrane permeability to Ca^{2+} is synchronous with the onset of membrane permeability to H^+, as indicated by loss of pattern N (fig. 2). The most probable explanation is that interaction of zona protein, in the case of the mouse ZP3, with the plasma membrane occurs through a receptor-mediated pathway which leads to a generalized permeability of both the plasma membrane and outer acrosomal membrane. The influx of Ca^{2+} into, and H^+ gradient loss from the acrosomal compartment appear to be parallel reactions. In mouse and possibly human sperm, the ZIR has the elements of a ligand receptor-mediated process, involving such participants in an intracellular signal transduction-mediated process as G_i proteins, protein kinase C, phospholipase C, and a possible receptor for the zona ligand that triggers the reaction. Precisely how these elements interact is as yet unclear, but the results of this study and some preliminary work carried out in our laboratory with the human system indicate that the endpoint of the process is free access of extracellular cations through the plasma and outer acrosomal membranes, allowing their entry into the acrosomal compartment.

References

Byrd, J.; Wolf, D.P.: Acrosomal status in fresh and capacitated human ejaculated sperm. Biol. Reprod. *34:* 859–869 (1986).

Cross, N.L.; Morales, P.; Overstreet, J.W.; Hanson, F.W.: Two simple methods for detecting acrosome-reacted human sperm. Gamete Res. *15:* 213–226 (1986).

Endo, Y.; Lee, M.A.; Kopf, G.S.: Characterization of an islet-activating protein-sensitive site in mouse sperm that is involved in the zona pellucida-induced acrosome reaction. Dev. Biol. *129:* 12–24 (1988).

Grynkiewicz, G.; Poenie, M.; Tsien, R.Y.: A new generation of Ca^{2+} indicators with greatly improved fluorescence properties. J. Biol. Chem. *260:* 3440–3450 (1985).

Lee, M.A.; Storey, B.T.: Evidence for plasma membrane impermeability to small ions in acrosome intact mouse spermatozoa bound to mouse zonae pellucidae, using an aminoacridine fluorescent pH probe: Time course of the zona-induced acrosome reaction monitored by both chlortetracycline and pH probe florescences. Biol. Reprod. *33:* 235–246 (1985).

Lee, M.A.; Storey, B.T.: Endpoint of first stage of zona pellucida-induced acrosome reaction in mouse spermatozoa characterized by acrosomal H^+ and C^{2+} permeability: Population and single cell kinetics. Gamete Res. *24:* 303–326 (1989).

Overstreet, J.W.; Yanagimachi, R.; Katz, D.F.; Hayashi, K.; Hanson, F.W.: Penetration of human spermatozoa into the human zona pellucida and the zona-free hamster egg: A study of fertile donors and infertile patients. Fertil. Steril. *33:* 534–551 (1980).

Saling, P.M.; Storey, B.T.: Mouse gamete interactions during fertilization in vitro. Chlortetracycline as a fluorescent probe for the mouse sperm acrosome reaction. J. Cell Biol. *83:* 544–555 (1979).

Ward, C.R.; Storey, B.T.: Determination of the time course of the capacitation in mouse spermatozoa using a chlortetracycline fluorescence assay. Dev. Biol. *104:* 287–296 (1984).

Wolf, D.P.; Boldt, J.; Byrd, W.; Bechtol, K.B.: Acrosomal status evaluation in human ejaculated sperm with monoclonal antibodies. Biol. Reprod. *32:* 1157–1165 (1985).

Michael A. Lee, The Cooper Institute for IVF, P.C., 8002 Greentree Commens, Marlton, NJ 08053 (USA)

Colpi GM, Pozza D (eds): Diagnosing Male Infertility.
Prog Reprod Biol Med. Basel, Karger, 1992, vol 15, pp 86–93

Purification of a Sperm Motility Inhibiting Factor from Human Peritoneal Fluids

G. Soldati, A. Piffaretti-Yanez, M. Balerna, U. Eppenberger

Andrology Laboratory, Gynaecological Endocrinology Unit, 'La Carità' Hospital, Locarno, Switzerland

The human peritoneal fluid (PF) microenvironment has been recognized as a dynamic zone of gamete interaction, sensitive to the influence of pathologic conditions and possessing constituents capable of having impact on reproductive processess. Several components of the PF have already been quantified and analyzed from a functional point of view. The relatively high total protein content of PF (55–65% of the serum value) accentuates its origin as an exudation product, where proteins with a low molecular weight pass through the capillary wall more easily than those with a high molecular weight [1]. A similar relationship has been demonstrated for the proteins in follicular fluid (FF) [2], and a striking similarity exists between FF:serum ratios [3] and PF:serum ratios [1]. Furthermore, prostaglandins [4], plasminogen activator [5], progestin-dependent endometrial protein [6], epidermal growth factor [7] and macrophages [8] producing interleukins and TNF [9] have been detected in PFs. Lipid components are also present in very different concentration from seminal plasma [10]. A role in PF lipids transfer was postulated to induce capacitation [11].

Among all constituents, four groups of substances have been analyzed: (a) substances capable of peroxidizing sperm plasma membrane lipids [12–14]; (b) prostaglandins and other substances produced by macrophages [15–18]; (c) macrophages and their direct contact with spermatozoa [19, 20], and (d) albumin and constituents directly bound or transported by the molecule [21–25].

Material and Methods

HPLC Separations. Pooled PF obtained from women participating in our GIFT/IVF-ET program was loaded on a Protein Pak 300 (MW sieving) HPLC column, the eluate collected (1-min fractions), dialyzed against bidest water (3.5 MWCO dialysis membranes), lyophilized and tested on sperm (see below). One milliliter of the inhibiting fraction was then loaded on an SP-MA7S HPLC column, the eluate collected in 4 fractions and tested on sperm. The SP-purified, active fraction was subsequently separated on a DEAE-5PW HPLC column and 5 fractions tested on sperm.

Sperm Contact Test. The fractions were tested on human sperm as described by Soldati et al. [26]. Briefly, 25 µl of motile-rich sperm suspension were mixed with 100 µl of PF fraction or with BWW-Hepes-0.3% HSA (pH 7.4) as a control. Three objective assessments of the percentage of motile spermatozoa were performed by the MEP technique at time 0, 2.5 and 5 h after mixing.

Results

Previous experiments using LPC (Sephacryl S-300) showed that a particular fraction of PF pool was able to completely immobilize sperm in 2.5 h as showed in figure 1. At the same time, sperm retained their viability [33]. HPLC experiments (MW sieving separations) gave immobilization times of 5 h; this time was reduced to 2.5 and 1 h after further purification of the active fraction on SP- and DEAE columns, respectively. These three separation steps are showed in figure 2 with the respective 'immobilizing fraction' in dashed lines. Each purification step consisted in further separate the already active fraction and testing it on sperm motility. Figure 3 shows the test of the 3 inhibiting fractions on the same population of spermatozoa submitted to a swim-up procedure. The control was in 0.3% HSA. The integrated 'immobilizing chromatography areas'/protein concentration ratios allowed to calculate a purification factor of 3 times for SP and 20 times for DEAE separations (table 1). This factor was found to be present in all of the PFs tested, its heat-stable and nondialyzable in 3.5 MW cut-off membranes (data not shown).

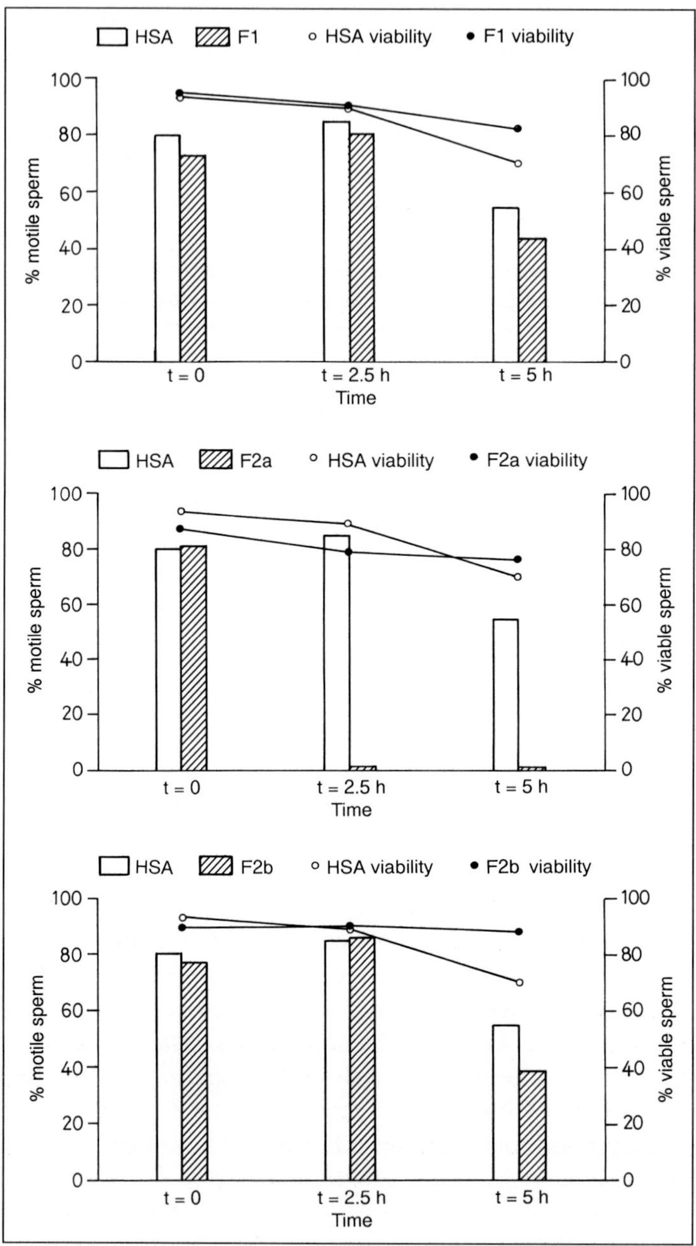

Fig. 1. Separation of a pooled PF on S-300 Sephacryl column in three distinct fractions and their respective effect on sperm motility at 0, 2.5 and 5 h.

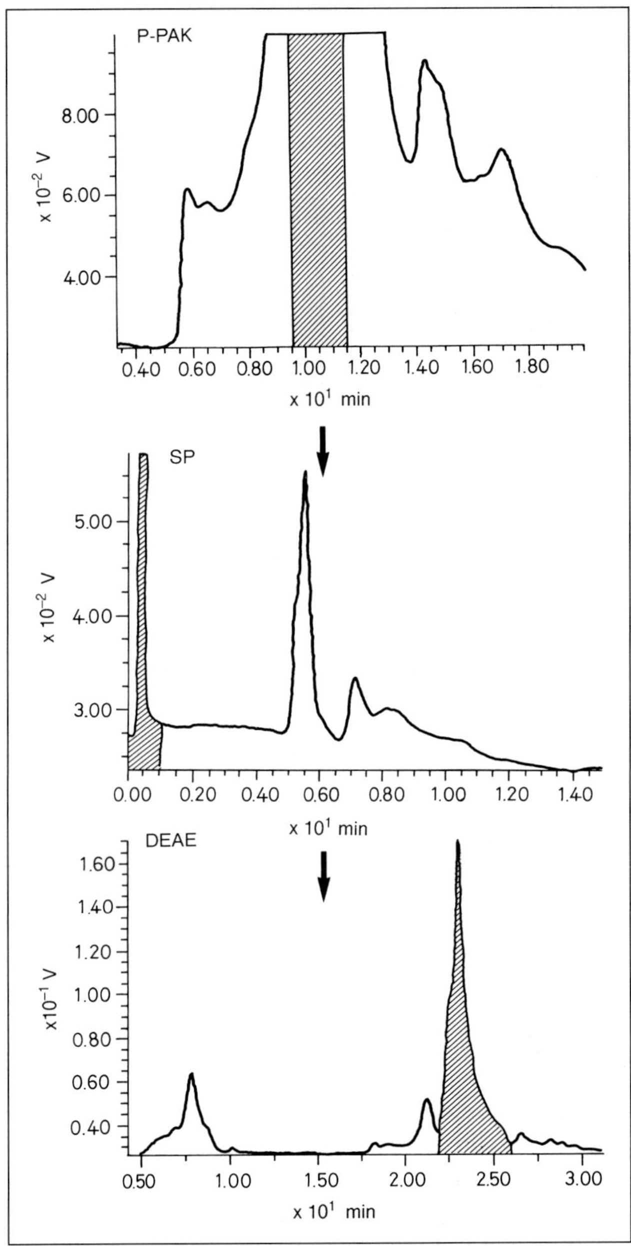

Fig. 2. Consecutive separations of a pooled PF on molecular sieve, SP and DEAE HPLC columns. The sperm-inhibiting fraction is highlighted.

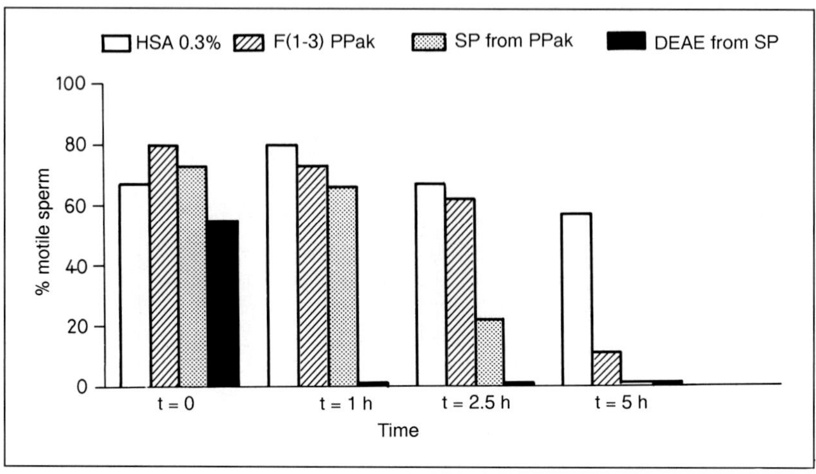

Fig. 3. Test on spermatozoa of the inhibiting fractions separated consecutively on three HPLC columns. MEP are taken at 0, 1, 2.5 and 5 h.

Table 1. Calculation of the purification factor of the immobilizing activity

	Activity/mg protein	Purification factor
Protein Pak 300	0.246	1
SP-MA7S	0.747	3.03
DEAE-5PW	5.03	20.45

$$\text{Factor} = \frac{1/\text{activity area}}{\text{protein conc.}}.$$

Discussion

No reports have appeared on the separation of PFs in view of testing its fractions on sperm motility, but several authors have reported the effects of FF fractions mainly on the acrosome reaction [27, 28]. Yanagimachi [29] postulated the presence in separated FFs of a low molecular weight factor which could induce the motility (sperm stimulating factor). Albumin has already been suspected to act on sperm cells as inducer of acrosome reaction [28], and capacitation via substances carried by the molecule [21] or acting directly on sperm membranes [23]. The importance of lipids bound to albumin has been emphasized by Lui and Meizel [25] in relation to acrosome reaction and by Davis and Byrne [22, 30, 31] in relation to lipid interaction with sperm plasma membrane.

These findings are indirectly confirmed by our study. In fact, in the third separation step (DEAE) we were able to sort a purified fraction in which albumin was the prominent, if not the only, protein present. The immobilizing fractions were all dialyzed and lyophilized and we could exclude a peroxidizing effect of these fractions on the sperm membranes (data not shown). We postulate that treatment of separated PF albumin in strong dialysis conditions could remove or substitute some substances (free fatty acids?) [32] and bind other substances which are normally not dangerous to spermatozoa. Entering in contact with sperm membranes these substances carried by albumin could render spermatozoa immotile.

References

1   Bouckaert, P.X.J.M.; Evers, J.L.H.; Doesburg, W:H.; Schellekens, L.A.; Brombacher, P.H.; Rolland, R.: Patterns of changes in proteins in the peritoneal fluid of women during the periovulatory phase of the menstrual cycle. J. Reprod. Fertil. *77:* 329–336 (1986).

2   Andersen, M.M.; Kroll, J.; Byskov, A.G.; Faber, M.: Protein composition in the fluid of individual bovine follicles. J. Reprod. Fertil. *48:* 109–118 (1976).

3   Edwards, R.G.: Follicular fluid. J. Reprod. Fertil. *37:* 189–219 (1974).

4   Halme, J.; Hammond, M.J.; Syrop, C.H.: Peritoneal macrophages modulate human granulosa-luteal cell progesterone production. J. Clin. Endocrinol. Metab. *16:* 912–917 (1985).

5   Batzofin, J.H.; Holmas, S.D.; Gibbons, W.E.; Buttram, V.C.: Peritoneal fluid plasminogen activator activity in endometriosis and pelvic adhesive disease. Fertil. Steril. *44:* 277–279 (1985).

6   Joshi, S:G.; Zamah, N.M.; Raikar, R.S.; Buttram, V.C.; Henriques, E.S.; Gordon, M.: Serum and peritoneal fluid proteins in women with and without endometriosis. Fertil. Steril. *46:* 1077–1081 (1986).

7   DeLeon, F.D.; Vijayakumar, R.; Brown, M.; Rao, C.V.; Yussman, M.A.; Schultz, G:: Peritoneal fluid volume, estrogen, progesterone, prostaglandin, and epidermal growth factor concentrations in patients with and without endometriosis. Obstet. Gynecol. *68:* 189–195 (1986).

8   Halme, J.; Becker, S.; Wing, R.: Accentuated cyclic activation of peritoneal macrophages in patients with endometriosis. Am. J. Obstet. Gynecol. *148:* 85 (1984).

9   Eisermann, J.; Gast, M.J.; Pineda, J.; Odem, R.R.; Collins, J.L.: Tumor necrosis factor in peritoneal fluid of women undergoing laparoscopic surgery. Fertil. Steril. *50:* 573–579 (1988).

10 Vignon, F.; Cranz, C.; Robillart, I.; Montagnon, M.; Clavert, A.; Pinget, M.: Etude comparative de la composition lipidique du liquide séminal et du liquide péritonéal ovulatoire dans l'espèce humaine. J. Gynecol. Obstet. Biol. Reprod. *18:* 459–462 (1989).

11 Davis, B.K.: Timing of fertilization in mammals: sperm cholesterol/phospholipid ratio as a determinant of the capacitation interval. Proc. Natl. Acad. Sci. USA *78:* 7560–7563 (1981).

12 Jones, R.; Mann, T.; Sherins, R.: Peroxidative breakdown of phospholipids in human spermatozoa, spermicidal properties of fatty acid peroxides, and protective action of seminal plasma. Fertil. Steril. *31:* 531–537 (1979).

13 Jones, R.; Mann, T.: Lipid peroxides in spermatozoa; formation, role of plasmalogen, and physiological significance. Proc. R. Soc. Lond. B *193:* 317–333 (1976).

14 Jones, R.; Mann, T.: Damage to ram spermatozoa by peroxidation of endogenous phospholipids. J. Reprod. Fertil. *50:* 261–268 (1977).

15 Cohen, M.S.; Colin, M.J.; Golimbu, M.; Hotchkiss, R.S.: The effects of prostaglandins on sperm motility. Fertil. Steril. *28:* 78–85 (1977).

16 Didolkar, A.K.; Roychowdhury, D.: Effects of prostaglandins E-1, F-1a and F-2a on human sperm motility. Andrologia *12:* 135–140 (1980).

17 Colon, J.M.; Ginsburg, F.; Lessing, J.B., et al.: The effect of relaxin and prostaglandin E-2 on the motility of human spermatozoa. Fertil. Steril. *46:* 1133–1139 (1986).

18 Hill, J.A.; Haimovici, F.; Politch, J.A.; Anderson, D.J.: Effect of soluble products of activated lymphocytes and macrophages (lymphokines and monokines) on human sperm motion parameters. Fertil. Steril. *47:* 460–465 (1987).

19 Muscato, J.J.; Haney, A.F.; Weinberg, J.B.: Sperm phagocytosis by human peritoneal macrophages: A possible cause of infertility in endometriosis. Am. J. Obstet. Gynecol. *144:* 503–510 (1982).

20 London, S:N.; Haney, A.F.; Weinberg, J.B.: Macrophages and infertility: enhancement of human macrophage-mediated sperm killing by antisperm antibodies. Fertil. Steril. *43:* 274–278 (1985).

21 Davis, B.K.; Byrne, R.; Bedigian, K.: Studies on the mechanism of capacitation: albumin-mediated changes in plasma membrane lipids during in vitro incubation of rat sperm cells. Proc. Natl. Acad. Sci. USA *77:* 1546–1550 (1980).

22 Davis, B.K.: Interaction of lipids with the plasma membrane of sperm cells. I. The antifertilization action of cholesterol. Arch. Androl. *5:* 249–254 (1980).

23 Dow, M.P.D.; Bavister, B.D.: Direct contact is required between serum albumin and hamster spermatozoa for capacitation in vitro. Gamete Res. *23:* 171–180 (1989).

24 Siegel, I.; Dudkiewicz, A.B.; Friberg, J.; Suarez, M.; Gleicher, N.: Inhibition of sperm motility and agglutination of sperm cells by free fatty acids in whole semen. Fertil. Steril. *45:* 273–279 (1986).

25 Lui, C.W.; Meizel, S.: Biochemical studies of the in vitro acrosome reaction inducing activity of bovine serum albumin. Differentiation *9:* 59–66 (1977).

26 Soldati, G.; Marchini, M.; Piffaretti-Yanez, A.; Luerti, M.; Campana, A.; Balerna, M.: Effect of peritoneal fluid on sperm motility and velocity distribution using objective measurements. Fertil. Steril. *52:* 113–119 (1989).

27 Suarez, S.S.; Wolf, D.P.; Meizel, S.: Induction of the acrosome reaction in human spermatozoa by a fraction of human follicular fluid. Gamete Res. *14:* 107–121 (1986).

28   Lui, C.W.; Cornett, L.E.; Meizel, S.: Identification of the bovine follicular fluid protein involved in the in vitro induction of the hamster sperm acrosome reaction. Biol. Reprod. *17:* 34–41 (1977).

29   Yanagimachi, R.: In vitro acrosome reaction and capacitation of golden hamster spermatozoa by bovine follicular fluid and its fractions (abstract). J. Exp. Zool. *170:* 269–280 (1970).

30   Davis, B.K.; Byrne, R.: Interaction of lipids with the plasma membrane of sperm cells. II. Evidence of a membrane thermotropic transition. Arch. Androl. *5:* 255–261 (1980).

31   Davis, B.K.; Byrne, R.: Interaction of lipids with the plasma membrane of sperm cells. III. Antifusigenic effect by phosphatidylserine. Arch. Androl. *5:* 263–266 (1980).

32   Hong, C.Y.; Shieh, C.C.; Wu, P.; Huang, J.J.; Chiang, B.N.: Effect of phosphatidylcholine, lysophosphatidylcholine, arachidonic acid and docosahexaenoic acid on the motility of human sperm. Int. J. Androl. *9:* 118–122 (1986).

33   Soldati, G.; Piffaretti-Yanez, A.; Balerna, M.: Effect of peritoneal fluid on sperm motility. Proc. Tri-National Meeting on Fertility and Sterility, Salzburg, 1989. In press.

Dr. G. Soldati, Andrology Laboratory, Gynecological Endocrinology Unit,
'La Carità' Hospital, CH–6600 Locarno (Switzerland)

Colpi GM, Pozza D (eds): Diagnosing Male Infertility.
Prog Reprod Biol Med. Basel, Karger, 1992, vol 15, pp 94–97

Is *Trichomonas vaginalis* a Cause of Male Infertility?

*M.S. Bornman, L. Grobler, D. Boomker, M.F. Mahomed,
G.W. Schulenburg, S. Reif, H.H. Crewe-Brown*

Andrology Laboratory, Departments of Urology and Microbiology,
Medunsa, South Africa

Trichomonas vaginalis was first described by Donné in 1836. The organism is a flagellate protozoan and the only one of three trichomonads which causes clinical disease in man [1]. *T. vaginalis* most frequently colonizes the vagina and cervix of women and the anterior urethra of the male sexual partners [2]. The disease is rare in prepubertal girls but common in sexually active men and women [2] with the highest prevalence in the age group 15–40 years [1]. Trichomoniasis may be the most common sexually transmitted disease [1].

There is little unanimity regarding *T. vaginalis* infection in the male [3]. The majority of *T. vaginalis* infections in men are asymptomatic [1, 3, 4], but some feel that *T. vaginalis* may be a significant cause of urological complications [5–7]. Little is known about the natural history of untreated *T. vaginalis* in males, except that the organism tends to disappear with time in some, while persisting in others, especially those with urethral stenosis [2]. These men may of course contribute to the spread of the disease.

Although it has been suggested by the studies of Tuttle et al. [8] that there is a possibility that *T. vaginalis* infections might impair fertility, clinical observations indicate that undiagnosed *T. vaginalis* infections rarely, if ever, cause male infertility [4, 9]. In this study we determined the incidence of asymptomatic carriage of *T. vaginalis* in sexually active males at an Infertility Clinic and studied the effect of infection on semen parameters.

Patients and Methods

Semen and first voided urine samples (urethral urine) were obtained from 396 males attending the Andrology Clinic of the Department of Urology at Medunsa between February and November 1989. Samples were collected at the clinic and maintained at $\pm 37\,°C$ during transport to the laboratory.

Ten milliliters of urine was centrifuged and the pellet used for direct microscopy, culture and an acridine orange (AO) stain. The wet mount was screened for the presence of motile organisms. Smears for AO staining were made on clean slides, air-dried and processed within 24 h [10]. AO is a substance which differentially stains DNA and RNA, and absorption of the stain is not dependent upon viability of the organism. The *T. vaginalis* nucleus characteristically stains green, whereas the rest of the organism is brick colored [10].

Diamond's medium was used for culture [11] and after inoculation the medium was incubated at $37\,°C$ in an atmosphere of 5–10% CO_2. Samples of the culture were examined by wet preparation after 18 h incubation and then daily thereafter for 7 days for the presence of *T. vaginalis* before regarding them as being negative.

Semen samples obtained by masturbation were collected at the clinic after a requested period of 3 days' sexual abstinence. After liquefaction, uncentrifuged semen samples were examined for *T. vaginalis* as above. Semen analyses were performed by the methods and standards of the WHO [12] and normal sperm morphology assessed by the criteria of Van Zyl et al. [13].

Results

T. vaginalis was demonstrated in 64 patients (16.2%). Of these cases, direct microscopy was positive in 2 (3%), AO staining in 14 (21.9%) and semen culture in 55 (86%) of cases. Thirty-seven (57.8%) of the urine samples were positive on microscopy, 43 (67.2%) on AO stain and 47 (73.4%) on culture. Nine (14%) cases, who were negative on semen culture, were positively diagnosed on AO staining of urine. Therefore, the combination of semen culture and acridine orange staining on urine was positive in all cases. Semen parameters of positive cases were not significantly different from those without infection.

Discussion

T. vaginalis was demonstrated in 64 patients (16.16%). Our cases were asymptomatic and therefore the asymptomatic carriage rate was 16%. The prevalence in control male populations ranges from 0 in asymptomatic men to 18% in men with gonococcal urethritis with a median prevalence in different studies of control male populations of 2% [1]. Semen parameters of positive cases were not significantly different from those without infection and it would appear that the presence of these trichomonads had no adverse effect on freshly ejaculated semen.

Reports on the prevalence of *T. vaginalis* are dependent upon many factors including the composition of the study group and the type of investigations used. The infection rate recorded by direct microscopy was very low in semen (3%) but high in urine (57.8%). The higher rate in the urethral urine might well be the result of the organism colonizing the anterior urethra from where it is washed out during micturition. The use of an early-morning urethral specimen has been suggested if parasites are scanty [1]. Our findings are in agreement with those of Daly et al. [4] who were rarely able to demonstrate trichomonads on direct spermatozoal microscopy. Specific care should be taken not to allow samples to cool to room temperature as trichomonads then rapidly lose their motility [1]. Since direct microscopy of unstained specimens is time consuming and detects only 3.6% of culture-positive cases, the use of other methods with a higher rate of positivity would be more appropriate.

AO staining of semen and urine yielded positive results in 22 and 67% of cases respectively. AO staining is a reliable method for the routine diagnosis of *T. vaginalis* in women [14]. Although the infection rate is lower than recorded by culture, the rapidity, ease and reliability of AO staining justify its use in the routine laboratory for the diagnosis of *T. vaginalis.*

It seems that special efforts should be made to recover *T. vaginalis* from clinical material. *T. vaginalis* grows best in cultures at 37 °C and under anaerobic conditions [1]. Occasional reports 'incorporating *T. vaginalis* for the sake of completeness should be interpreted cautiously' [1]. The highest percentage of positives in this study were found by culturing semen (86%) and urine (73%).

From our results it would seem that culture yielded more positive results than direct methods. The combination of semen culture and AO staining of urine was positive in all cases and is the diagnostic approach of choice in cases where *T. vaginalis* is suspected.

References

1   Krieger, J.N.: Urologic aspects of trichomoniasis. Invest. Urol. *6:* 411–417 (1981).
2   Von Lichtenberg, F.; Lehman, J.S.: Parasitic diseases and the genitourinary system; in Walsh, P.C.; Gittes, R.F.; Perlmutter, A.D.; Stamey, T.A. (eds): Campbell's Urology, pp. 1011–1015 (Saunders, London 1986).

3   Lanceley, F.: *Trichomonas vaginalis* infections in the male. Br. J. Venereol. Dis. *29:* 213–217 (1953).
4   Daly, J.J.; Sherman, J.K.; Green, L.; Hostetler, T.L.: Survival of *Trichomonas vaginalis* in human semen. Genitourin. Med. *65:* 106–108 (1989).
5   Jirovec, O.; Petru, M.: *Trichomonas vaginalis* and trichomoniasis. Adv. Parasitol. *6:* 177 (1968).
6   Honigberg, B.M.: Trichomonads of importance in human medicine; in Kreier, J.P. (ed.): Parasitic Protozoa (Academic Press, New York 1978).
7   Fourts, A.C.; Kraus, S.J.: *Trichomonas vaginalis:* Reevaluation of its clinical presentation and laboratory diagnosis. J. Infect. Dis. *141:* 137 (1980).
8   Tuttle, J.P.; Holbrook, T.W.; Derrick, F.C.: Interference of human spermatozoal motility and *T. vaginalis*. J. Urol. *118:* 1024 (1977).
9   Fowler, J.E.; Kessler, R.: Genital tract infection; in Lipshultz, L.I.; Howards, S.S. (eds): Infertility in the Male, pp. 283–298 (Churchill Livingstone, New York 1983).
10 Hipp, S.S.; Gaafar, M.W.: Screening for *Trichomonas vaginalis* infection by use of acridine orange fluorescent microscopy. Sex. Transm. Dis. *6:* 235–238 (1979).
11 Diamond, L.S.: The establishment of various trichomonads of animal and man in axenic cultures. J. Parasitol. *43:* 488–490 (1957).
12 WHO Manual for the Examination of Semen and Semen-Cervical Mucus Interaction (Cambridge University Press, Cambridge 1987).
13 Van Zyl, J.A.; Menkveld, R.; van W. Kotze, T.J.; Van Niekerk, W.A.: The importance of a spermiogram that meets the requirements of international standards and the most important factors that influence semen parameters. Proc. 17th Int. Soc. Urology, Paris 1976, p. 263 (Diffusion Doins, Paris 1976).
14 Mason, P.R.; Super, H.; Fripp, P.J.: Comparison of four techniques for the routine diagnosis of *Trichomonas vaginalis* infection. J. Clin. Pathol. *29:* 154–157 (1976).

Dr. M.S. Bornman, Andrology Laboratory, Departments of Urology and Microbiology, Box 242, Medunsa 0204 (South Africa)

Colpi GM, Pozza D (eds): Diagnosing Male Infertility.
Prog Reprod Biol Med. Basel, Karger, 1992, vol 15, pp 98–102

Detached Ciliary Tufts in Semen: An Indication of Impaired Epididymal Function?

*M.S. Bornman, D. du Toit, M.H. Fourie, C. de Villiers-Smith,
E.L. Kok, D.J. du Plessis*

Spermatology Laboratory, Department of Urology, University of Pretoria,
South Africa

Detached ciliary tufts are the shedded distal portion of ciliated epithelial cells [1]. These structures have been observed in pulmonary cytological material [1], nasal secretions [2], cervicovaginal smears [3, 4], in ovarian cyst aspirates (occasionally) [5] and in fluid aspirated from the pouch of Douglas [6]. These tufts were recently observed in the semen of some men presenting with infertility [7].

In the male genital tract, cilia are found in the ductuli efferentes, on the cell surfaces of the principal cells of the epididymis and on the epithelium of the vas deferens [8]. Motile cilia facilitate the movement of sperm in the ductuli efferentes, but the cilia of the epididymal duct are not regarded as true cilia, not possessing a capacity for purposeful movement [9].

During migration through the epididymis, sperm acquire the capacity for progressive movement [10]. Transit through the epididymis is accompanied by a decrease in sperm [ATP] [11]. Seminal plasma α-glucosidase activity has been proposed as a marker of both epididymal function and patency [12]. The aims of this study were to report on the clinical features of these patients and their epididymal function by comparing sperm [ATP] and seminal α-glucosidase activity in a pilot group with DCTs and a control group.

Patients and Methods

Eighty-one patients who were evaluated over a 3-year period for infertility at the Spermatology Laboratory of the Department of Urology, H.F. Verwoerd Hospital, Pretoria, were found to have DCTs in their semen samples. A detailed history was taken and each patient clinically examined with particular attention taken to testicular volume and consistency. Testes measuring < 4 cm longitudinally were regarded as small [13]. Semen was collected through masturbation at the clinic and analyzed according to the methods and standards of the World Health Organization [14]. Seminal smears, including those from cases with severe oligo- and azoospermia, were air-dried, stained (using the Papanicolaou method) and assessed by a single observer using the Tygerberg criteria [15]. Serum follicle-stimulating hormone (FSH), serum luteinizing hormone (LH) and testosterone were measured using radioimmunoassay kits.

In another 10 patients who also presented with DCTs in semen, sperm ATP concentration and seminal α-glucosidase activity were studied and compared to a control group where no DCTs were observed. The [ATP]/106 sperm was determined bioluminometrically [16] and the total α-glucosidase activity per ejaculate spectrophotometrically [12]. Statistical analysis was done using the Mann-Whitney test.

Results

DCTs were observed in multiple semen samples obtained from these 81 patients. The mean age was 31 years (range 23–40 years). Almost 50% (42 patients) smoked and of these, 37 smoked > 10 cigarettes per day. The mean number of cigarettes smoked by the group was 22 per day. Twenty-seven patients (34%) had small testes with a soft consistency. Altogether 32 patients (40%) had varicoceles (10 of these varicoceles were only Doppler positive). Twelve patients had histories of urogenital surgery, including varicocelectomies, testicular biopsies and vasovasostomies, done prior to this study. An elevated serum FSH was found in 3 cases. The mean serum testosterone for patients with DCTs was 17.63 ± 5.62 nmol/l, as compared to 26.74 ± 8.89 nmol/l of an age- and time-matched control group (normal range 18–38 nmol/l).

Seventeen patients (21%) had either very severe oligozoospermia or azoospermia, and in another 27 patients (38%) the sperm densities were less than 20 million/ml. Thirty-seven (41%) were normozoospermic. Sperm morphological assessment was possible in 61 (75%) cases of the 81 cases. In these, sperm density, motility and forward progression were all relatively normal on several occasions. The most outstanding findings in these patients were a low percentage of normal sperm (mean 7%) and an increased frequency of cytoplasmic droplets (mean $6.13 \pm 4.21\%$; WHO limits 2 ± 2%). The incidence of amorphous heads was 89%, neck and tail abnormalities 25% and precursors 3%.

The ATP concentration in the DCT group was 43.33265 ± 31.8 fg/million sperm and 4.7502373 ± 2.4 fg/million sperm in the control group (p = 0.009). The α-glucosidase activity in the DCT group was 56.12 ± 39.5 IU/ejaculate compared to 93.79 ± 32.2 IU/ejaculate in the control group (p = 0.0412).

Discussion

Shedding of DCTs and their appearance in semen implies a pathological process as suggested by the impaired fertility in these cases. From the occurrence of a wide range of clinical features in these patients, it is evident that shedding is probably not the result of a single etiological agent. However, certain factors were common in the individuals of the group. The major features of patients with DCTs in semen were a high incidence of smoking, diminished testicular volume in 34%, relatively low serum testosterone, a 40% incidence of varicoceles and an increased percentage of cytoplasmic droplets.

Smoking may affect spermatogenesis and epididymal functional directly as a result of the concentration of nicotine in tissue, catecholamine release, or by impairing Leydig cell function [17, 18]. The high incidence of smoking and lowered serum testosterone levels found in these patients may not be coincidental.

A decrease in testicular volume is often associated with impaired spermatogenesis [13]. Although the etiology of the decreased volume is not clear, the lower testicular volume points to possible testicular pathology. Impaired epididymal function is often associated with testicular pathology, as in the case of a varicocele. During epididymal transit, the cytoplasmic droplet migrates along the middle piece of the sperm and is subsequently shed. An increased frequency of cytoplasmic droplets on ejaculated sperm has been reported to indicate impaired epididymal function [13], presumably irrespective of the etiology.

Cilia are present in the male genital tract in the ductuli efferentes, on the cell surface of the principal cells of the epididymis and on the epithelium of the vas deferens. Although the ductuli efferentes could be the origin of the DCTs, the transit time of 12 days through the epididymis [9] and the possibility of phagocytosis during epididymal transit [19] renders this highly unlikely. The only evidence of a possible vasal origin of the DCTs was found in 2 patients who had had vasovasostomies after previous vasectomies. Therefore, it is postulated that these DCTs were most probably shed from the epididymis.

Patients with DCTs had a significantly *higher* sperm [ATP] and lower α-glucosidase activity than those in the control group. During transit through the epididymis, sperm mature and acquire the capacity for forward movement [10], which is eventually necessary for the development of fertilizing ability [13]. Transit through the epididymis is accompanied by a decrease in sperm [ATP] [11, 20] and an increase in metabolic activity, which is a distinctive part of the maturation process [21]. The higher sperm [ATP] in patients with DCTs could therefore reflect a possible impaired epididymal function. The same might also apply to the lower α-glucosidase activity, which has been suggested as a marker of epididymal patency [22] and epididymal function [12, 23].

These results would seem to point to impaired epididymal function in patients with DCTs and further support the proposed epididymal origin of DCTs. It would also seem possible that DCTs observed in semen might give the clinician an indication of impaired epididymal function.

References

1   Papanicolaou, G.N.: Degenerative changes in ciliated cells exfoliating from the bronchial epithelium as a cytologic criterion in the diagnosis of diseases of the lung. N.Y. State J. Med. *56:* 2647–2650 (1956).
2   Hilding, A.: The common cold. Arch Otolaryngol. *12:* 133–150 (1930).
3   Hollander, D.H.; Gupta, P.K.: Detached ciliary tufts in cervicovaginal smears. Acta Cytol. *18:* 367–369 (1975).
4   Muller Kobold-Wolterbeek, A.C.; Beyer-Boon, M.E.: Ciliocytophthoria in cervical cytology. Acta Cytol. *19:* 89–91 (1975).
5   Marinaccio, G.; Pollice, L.; Rizzo G.; Prete, F.: Unusual ciliated epithelial cells from an ovarian cyst. Acta Cytol. *18:* 65–67 (1974).
6   Poropatich, C.; Ehya, H.: Detached ciliary tufts in pouch of Douglas fluid. Acta Cytol. *30:* 442–444 (1986).
7   Otto, B.S.; du Plessis, D.J.: Detached ciliary tufts in semen. S. Afr. Med. J. *72:* 441 (1987).
8   Bloom, W.; Fawcett, D.W.: Male reproductive system; in Bloom, W.; Fawcett, D.W. (eds.): A Textbook of Histiology, pp. 843–846 (Saunders, London 1975).
9   Lipshultz, L.I.; Howards, S.S.: Evaluation of the subfertile man; in Lipshultz, L.I.; Howards, S.S. (eds.): Infertility in the Male, pp. 127–128 (Churchill Livingstone, New York 1983).

10 Longo, F.J.: Fertilization, p. 5 (Chapman & Hall, London 1987).

11 Hoskins, D.D.: Adenine nucleotide mediation of fructolysis and motility in bovine epididymal spermatozoa. J. Biol. Chem. *248:* 1135 (1973).

12 Cooper, T.G.; Yeung, C., et al.: Epididymal markers in human infertility. J. Androl. *9:* 91–101 (1988).

13 Lipshultz, L.I.; Howards, S.S.: Evaluation of the subfertile man; in Lipshultz, L.I.; Howards, S.S. (eds.): Infertility in the Male, pp. 121–190 (Churchill Livingstone, New York 1983).

14 Belsey, M.A.; Eliasson, R., et al.: Laboratory manual for the examination of human semen and semen-cervical mucus interaction. WHO Special Programme (Press Concern, Singapore 1980).

15 Van Zyl, J.A.; Menkveld, KR.; van W. Kotze, T.J.; Van Niekerk, W.A.: The importance of a spermiogram that meets the requirements of international standards and the most important factors that influence semen parameters. Proc. 17th Int. Soc. Urology, Paris 1976, p. 263 (Diffusion Doins, Paris 1976).

16 Du Toit, D.: M. Med. Sc. thesis, University of the Orange Free State, Bloemfontein (1989).

17 Raboch, J.; Mellan, J.: Smoking and fertility. Br. J. Sex. Med. *2:* 35 (1975).

18 Shaarawy, M.; Mahmoud, K.Z.: Endocrine profile and semen characteristics in male smokers. Fertil. Steril. *38:* 255–257 (1982).

19 Mann, T.; Lutwak-Mann, C.: Male reproductive function and the composition of semen: general considerations; in Mann, T.; Lutwak-Mann, C. (eds.): Male Reproductive Function and Semen (Springer, New York 1984).

20 Frenkel, G.; Peterson, R.N.; Freund, M.: The role of adenine nucleotides and the effect of caffeine and dibutyryl cyclic AMP on the metabolism of guinea pig epididymal spermatozoa. Proc. Soc. Exp. Biol. Med. *144:* 420–425 (1973).

21 Courot, M.: Transport and maturation of spermatozoa in the epididymis of mammals; in Bollack, C.; Clavert, A. (eds.): Prog. Reprod. Biol. Med., vol 8 (Karger, Basel 1981).

22 Guerin, J.-F.; Ben Ali, H.; Rollet, J.; Souchier, C.; Czyba, J.-C.: α-Glucosidase as a specific epididymal enzyme marker: Its validity for the etiologic diagnosis of azoospermia. J. Androl. *7:* 156–162 (1986).

23 Viljoen, M.H.; Bornman, M.S.; van der Merwe, M.P.; du Plessis, D.J.: Alpha-glucosidase activity and sperm motility. Andrologia (in press).

Dr. M.S. Bornman, Spermatology Laboratory, Department of Urology, University of Pretoria, Private Bag X169, Pretoria 0001 (South Africa)

Colpi GM, Pozza D (eds): Diagnosing Male Infertility.
Prog Reprod Biol Med. Basel, Karger, 1992, vol 15, pp 103–108

Automated Analysis of Human Sperm Motility by Computer-Assisted Image Processing

Walter Krause, Gabriele Schönhärl

Department of Andrology, Center of Dermatology, University Clinics,
Philipps University, Marburg, FRG

The motility of human spermatozoa is one of the most important parameters of fertility prognosis and objectives for therapeutic procedures. Thus, it is necessary to determine this parameter by objective procedures in a reproducible manner. The methods used in the routine laboratory so far are imprecise and underlie subjective influences by the investigator or the technician.

Computer-assisted image processing (CAIP) offers the possibility to overcome these difficulties. A computer analyses the microscopic picture observed by a videocamera and counts the objects found in the given field. For the analysis of movement of cells, several frames with a given, very small time interval are analysed and the path of an object (a spermatozoon) is calculated. Thus, the ratio of motile cells and their motility parameters can be calculated. To exclude objects which are not spermatozoa from the measurement, a size threshold is used. All objects exceeding these in size are excluded. The other parameter used in the grey level. The computer is able to calculate a so-called grey image from the original image, which is based on the optical density of the objects seen. One has to adjust a threshold, and all objects with a grey level below this are excluded from the measurement. Lastly, the results of the measurement are printed and stored in the computer. They are available for further statistical analysis.

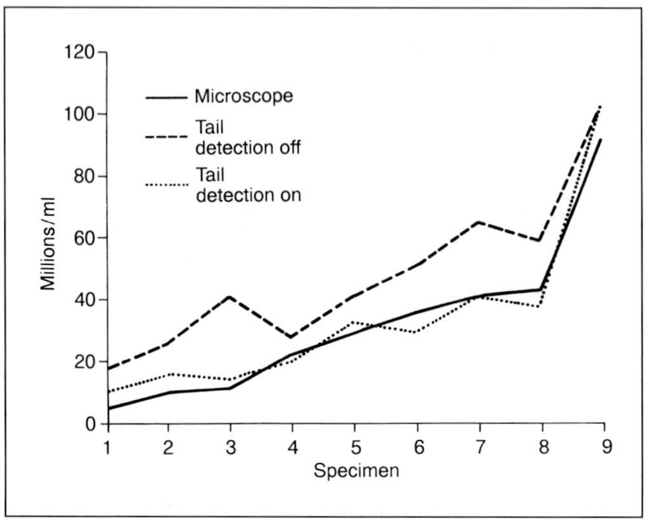

Fig. 1. Results of counting spermatozoa in several samples by the microscopic counting chamber (——), the SM-CMA (– – –) and the SM-CMA with 'tail detection' switched off (·····).

Methods

The system we are now using, the sperm motion analyser from Strömberg-Mika (SM-CMA), is of further help to identify cells as spermatozoon: the tail detection. After having finished a measurement, the system examines all objects held for a tail and excludes those without a tail-like structure from the measurement. Figure 1 gives the results of counting some semen samples. The results of counting are more invalid without tail detection. However, it is also clear from this figure that the results of counting are less valid when the cell count is low, i.e. in oligozoospermia. In such semen samples there are more cell debris and other particles, which give false results. The phenomenon is one of the reasons why the present systems are not very useful for use in the routine andrologic laboratory.

Results

When comparing the results of motility analysis by CAIP and the routinely assessed microscopic examinations, we observed the following correlations: immotile spermatozoa as counted with the microscope and those seen by the system show a fairly good correlation (fig. 2). It is not

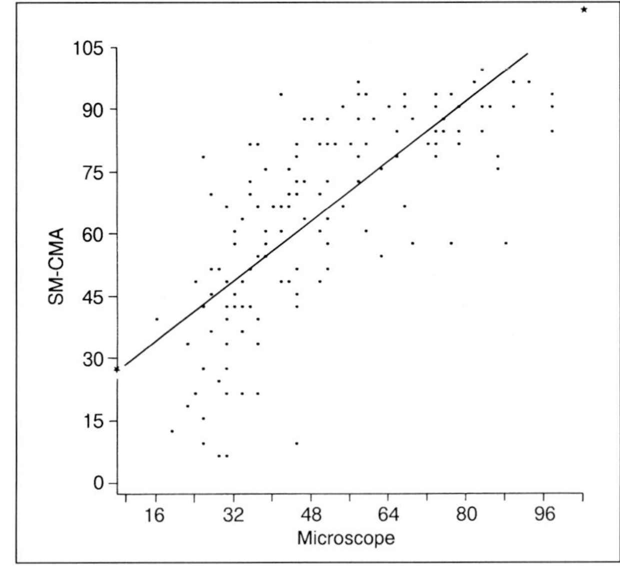

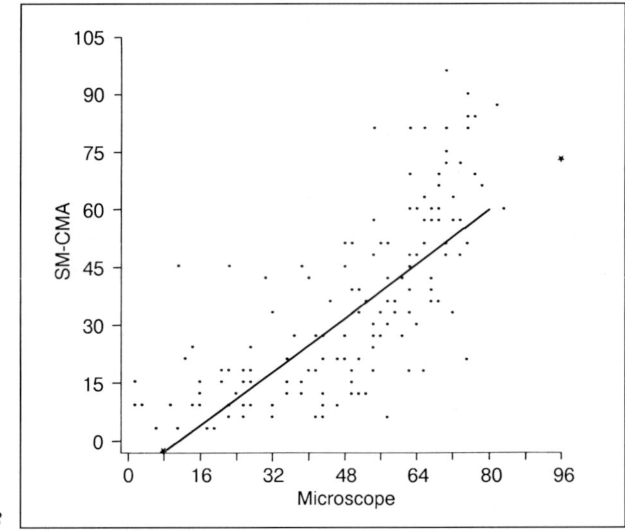

Fig. 2. Correlation between the number of immotile sperm estimated with the microscope and the SM-CMA. 148 cases plotted. Regression statistics. Correlation r = 0.73091. Intercept 19.52800, slope 0.87757.

Fig. 3. Correlation between the total number of motile sperm estimated with the microscope and the SM-CMA. 148 cases plotted. Regression statistics. Correlation r = 0.73045. Intercept – 7.12986, slope 0.86927.

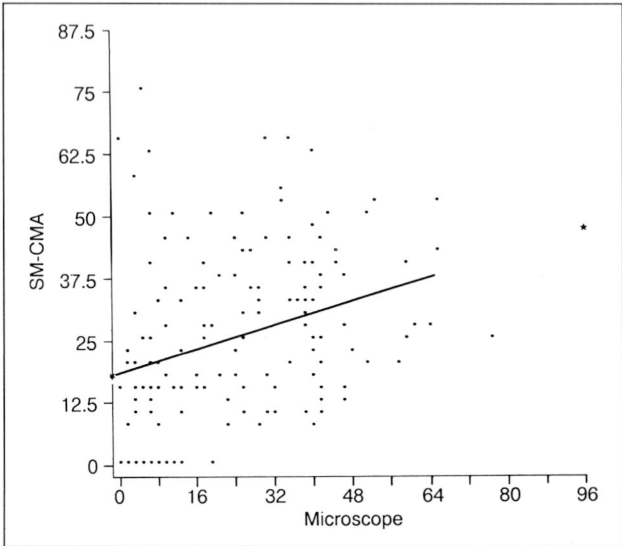

Fig. 4. Correlation between progressive motile sperm estimated with the microscope and the SM-CMA. 148 cases plotted. Regression statistics. Correlation r = 0.32753. Intercept 18.00649, slope 0.31718.

surprising, because the parameter 'tail detection' indeed enables the system to identify spermatozoa – both determinations recognize the same objects. Figure 3 shows that there is also a close correlation between the total number of motile cells determined by the microscope and those determined by the SM-CMA.

The correlation is low between the number of linear motile cells measured by the CAIP and that of progressive motile cells as judged in the microscope (fig. 4). The decision by the investigator 'this is a progressive motile cell, the other cell is not' is very imprecise.

When comparing local motility measured by both procedures (fig. 5), no correlation is visible, because the definition by the investigator is again imprecise. One has to be aware of the observation that, if one has a very low number of progressive motile cells, one tends to shift some local motile cells to the higher class of progressive motility. When correlating the number of progressive motile and the mean velocity of cells (VAP) as measured by the CAIP, there is a fairly good correlation (fig. 6). Spermatozoal velocity cannot be measured in the microscopic evaluation.

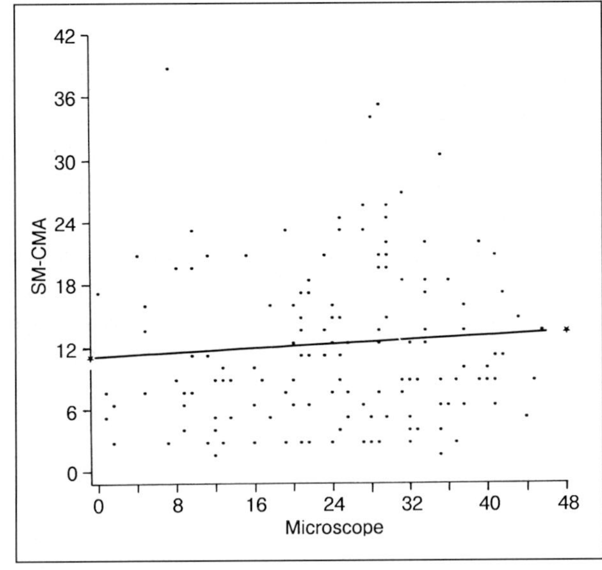

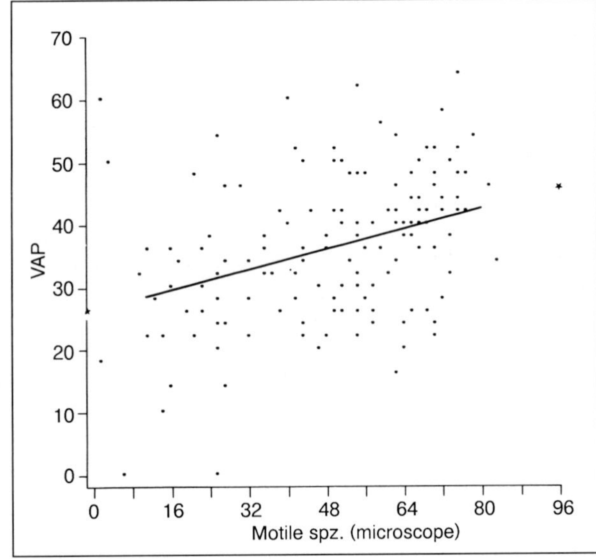

Fig. 5. Correlation between the number of local motile sperm estimated with the microscope and the SM-CMA. 148 cases plotted. Regression statistics. Correlation r = 0.08067. Intercept 10.22808, slope 0.05299.

Fig. 6. Correlation of mean sperm velocity (VAP) and motile cells. 148 cases plotted. Regression statistics. Correlation r = 0.36879. Intercept 25.28249, slope 0.21179.

Discussion

Several investigators published their experiences with CAIP systems for the determination of sperm motility and described little correlation to the microscopic determinations [1, 2, 3]. This is not due to false measurement by CAIP, but due to the subjective and irreproducible estimation of parameters by the microscope.

The CAIP offers the possibility of measuring new parameters having a much lower variation than those obtained by microscopic evaluation. They are possibly more suitable for the description of fertility prognosis, since they show a lower variation between different measurements in the same individual [4, 5].

References

1   Albe X, Auger J, Margulkes S, et al. : Measurement of spermatozoal motility over prolonged periods of time using image analysis. Comput Biomed Res 1988;21:276–288.
2   Auger J, Dadoune JP: Computerized sperm motility and application of sperm cryoconservation. Arch Androl 1988;20:103–112.
3   Mortimer D, Serres C, Mortimer ST, et al.: Influence of image sampling frequency on the perceived movement characteristics of progressively motile human spermatozoa. Gamet Res 1988;20:313–327.
4   Chan SY, Wang C, Ng M, et al.: Evaluation of computerized analysis of sperm movement characteristics. J Androl 1989;10:133–138.
5   Knuth UA, Kuhne J, Bals-Pratsch M, Nieschlag E: Intraindividual variation of sperm velocity, linearity, lateral head displacement and beat frequency in healthy volunteers. Andrologia 1988;20:243–248.

Dr. Walter Krause, Department of Andrology, Center of Dermatology, University Clinics, Philipps University, Deutschhausstrasse 9, D-W–3550 Marburg (FRG)

Colpi GM, Pozza D (eds): Diagnosing Male Infertility.
Prog Reprod Biol Med. Basel, Karger, 1992, vol 15, pp 109–113

Subtle Motility Abnormalities Diagnosed by Computer-Assisted Semen Analysis Not Found to Correlate with Infertile Males Better than Conventional Analysis

Jeffrey Chase[a], *Jerome H. Check*[a], *Mark Syrkin*[a], *Mike Lee*[a], *Jan Kozak*[b]

[a] The UMDNJ, Robert Wood Johnson Medical School at Camden, Cooper Hospital/University Medical Center, Department of Obstetrics/Gynecology, and [b] Division of Reproductive Endocrinology and Infertility and Division of Research, Camden, N.J., USA

One of the theoretical advantages of the computer-assisted semen analysis (CSA) is that by objectively isolating different motility factors some abnormalities may be noted by CSA that will correlate with male fertility that are not detected by routine semen analysis. Some data have been presented demonstrating that routine semen analysis (RSA) determined 50% of the male partners of a group of infertile couples that were found to have a subnormal RSA but the CSA was more sensitive detecting 80% abnormalities in the males [1].

However, the ability to detect a higher percentage of 'abnormal' males does not necessarily indicate that these males are truly subfertile. A recent study demonstrated no significant differences in the 6 month pregnancy rates of couples where a female factor was identified and corrected in the subset where the male partner had a normal motile density (83%) versus those males with subnormal levels (69%) [2]. Perhaps some CSA parameter(s) would be identified that could predict a fertile male even when the RSA was subnormal.

Not only has the RSA failed to adequately identify fertile males with subnormal parameters but the opposite has been found, i.e. males with normal RSA but who are subfertile. This was illustrated by demonstrating a high 6-month pregnancy rate after therapeutic donor insemination (TDI)

in a group of infertile couples failing to conceive despite the male seemingly normal by RSA and all female infertility factors corrected (and given at least 8 corrected cycles) [3].

A study was thus initiated to evaluate a group of infertile couples where a female factor was identified and believed to be correctable to determine if the CSA would better discriminate the fertile versus subfertile male.

Materials and Methods

A group of 405 couples with a minimum of 1 $^{1}/_{2}$ years of infertility were evaluated in this study. An additional requirement for inclusion was that a female factor(s) be identified and thought to be corrected. In this first group of 155 couples the semen analysis was performed manually, employing the Makler chamber for count and motility. The quality of motility and sperm morphology was estimated using the phase-contrast microscope. This group of tests is known as RSA. The other group consisted of 250 couples. The semen was evaluated by CellSoft (Cryo Resources, Ltd., New York, N.Y.) for a CSA; RSA was also performed. The group of parameters for CSA included: velocity, linearity, lateral head displacement, and beat/cross frequency, the normals were suggested by Cryo Resources, and correlates with lowest values obtained from our sperm donor specimens. The norms (parameters of normal males) for routine semen analysis were suggested by the World Health Organization (WHO) (table 1).

Quality of motility for RSA was judged on a scoring system of 1 to 4. The specimen was considered to be of subnormal quality if grade was 1 or 2. Grade 1 = minimal forward progression; grade 2 = poor to fair motility; grade 3 = good activity with linear forward progression; grade 4 = full activity with both fast and linear forward progression.

An RSA was considered subnormal if the motile density was below 10×10^6/ml or morphology was < 50% normal. A CSA was abnormal if any parameter fell below normal standards.

Table 1. Normal values for RSA and CSA

	Normal values
Count	$\geq 20 \times 10^6$/ml
Motility	$\geq 50\%$ grade 3 or 4
Morphology	$\geq 50\%$ normal forms
Velocity	$\geq 40\ \mu$m/s
Linearity	≥ 5.5
ALH mean	$\geq 1.8\ \mu$m
ALH max	$\geq 2.3\ \mu$m
Beat/cross frequency	≥ 13 Hz (1/s)

Results

The pregnancy rates according to RSA and CSA parameters are shown in tables 2-4. The data showed that 70% of the time the RSA and CSA correlated with each other as to whether it was normal or subnormal. Interestingly, the pregnancy rates were almost identical whether the semen analyses were normal or subnormal (60 and 62%, respectively).

Table 2. Pregnancy rates according to RSA and CSA parameters

	155 couples			250 couples		
	patients		pregnancy rates	patients		pregnancy rates
	n	%	%	n	%	%
RSA only subnormal	48/155	31	63	59/260	24	58
RSA only normal (CSA not performed)	107/155	69	85			
CSA only subnormal				16/250	6	31
RSA + CSA normal				55/250	22	60
RSA + CSA subnormal				120/250	48	62

Table 3. Computerized semen parameters vs. pregnancy rates for the group of 250, evaluated by RSA and CSA

	Number normal	Pregnant		Number subnormal	Pregnant	
		n	%		n	%
Count	86+131 = 217	131	60	18+15 = 33	15	45.5
Motility	41+56 = 97	56	58	63+90 = 153	90	59
Velocity	74+97 = 171	97	57	30+49 = 79	49	62
Linearity	76+106 = 182	106	58	28+40 = 68	40	59
ALH mean	74+107 = 181	107	59	30+39 = 69	39	57
Beat/cross	88+131 = 219	131	60	16+15 = 31	15	48
Morphology	70+87 = 157	87	55	34+59 = 93	59	63

Total number of pregnant couples = 146; total number of non-pregnant couples = 104.

Table 4. Comparison of semen variables in the group of couples who conceived and the group who did not (means ± SD)

Semen variables	Pregnant	Non-pregnant
Count × 10^6/ml	90.5 ± 95.4	84.5 ± 70.2
Motility, %	44.1 ± 20.2	44.5 ± 17.9
Velocity, μm/s	44.8 ± 9.2	44.1 ± 9.7
Linearity	6.2 ± 1.2	6.0 ± 0.9
ALH (mean), μm	2.1 ± 0.5	2.2 ± 0.6
Beat/cross frequency, 1/s	15.5 ± 1.8	15.3 ± 1.8

A priori, the theoretical advantage of the computer would be to identify an abnormal male factor that the naked eye could miss and might falsely consider a semen analysis normal. When there was a discordancy between RSA and CSA (30%) only 16/75 (21.3%) found the RSA normal but the CSA detected a 'subtle abnormality'; in fact, the majority of the discrepancies centered on the RSA subnormal but the CSA normal 59/75 (78.7%).

Though the group with the exclusive abnormality in the CSA did have the lowest pregnancy rate at 31% (5/16), the numbers are too small to show statistical significance. The fact that the larger group with both subnormal RSA and CSA had a 62% pregnancy rate (74/120) suggests that it would be highly unlikely that the group with RSA normal but CSA subnormal would continue to demonstrate the lowest pregnancy rates.

Evaluating the CSA slighly differently, i.e. considering 2 or more parameters abnormal, 61/104 (58.7%) of non-pregnant cases were abnormal compared to 87/146 (59.6%) of the pregnant cases. Finally, no single parameter of the CSA that predicted a male with subfertility was found.

Discussion

Some of the potential physical pitfalls present in various CSA analyses have been described by Vantman et al. [4]. Most instruments are very expensive. Thus, a very important question is whether these instruments should be purchased by the average clinician practicing infertility or restricted just to research uses. If studies would indicate that CSA as opposed to RSA could better predict the subnormal male then the expensive purchase would be justified.

Unfortunately, CSA proved no more effective than RSA in distinguishing the fertile from the subfertile male and, in fact, except possibly for extremely poor specimens, neither of these tests were useful as fertility potential tests. These data emphasize the need to find other methods of evaluating the semen analysis. Some preliminary data indicates that evaluating motion characteristics of sperm after being washed (capacitated) may prove useful [5]. CSA may be used for the latter. Thus, the main role of CSA in andrology should be limited to research.

References

1   Young PE: Measurement of more sensitive parameters of male fertility by computer assisted semen analysis. Fertility Institute Medical Group of San Diego Inc. Pacific Coast OB/GYN Soc, 1988.
2   Check JH, Epstein R, Nowroozi K, Shanis B, Wu CH, Bollendorf A: The hypoosmotic swelling test as a useful adjunct to the semen analysis to predict fertility potential. Fertil Steril 1989;52:159–161.
3   Check JH, Framroze A, Liss J, Bollendorf A: Therapeutic insemination by donor achieves high pregnancy rates in infertile couples regardless of the males motile sperm density. Arch Androl, in press.
4   Vantman D, Kokoulis G, Dennison L, Zinaman M, Sherins R: Computer assisted semen analysis: Evaluation of method and assessment of the influence of sperm concentration on linear velocity determination. Fertil Steril 1988;49:510–515.
5   Banks S, Koukoulis A, Dennison L, Sherins RJ, Vantman D: Curvilinear velocity assessed during capacitation distinguishes sperm from infertile men not recognized by velocity measurements in seminal plasma: IVth Int Congr Andrology, Florence, 1989, p 114.

Jerome H. Check, MD, 7447 Old York Road, Melrose Park, PA 19126 (USA)

Colpi GM, Pozza D (eds): Diagnosing Male Infertility.
Prog Reprod Biol Med. Basel, Karger, 1992, vol 15, pp 114–118

Influence of the Counting Chamber during Computer-Assisted Semen Analysis with the SM Motion Analyzer

U. Schneider, W.G. Gehring, O. Buurman

Institute for Reproductive Medicine, Springe, FRG

The assessment of sperm concentration is based on counting the number of spermatozoa with the help of a counting chamber. A classical haemocytometer or modified counting chamber (Makler, Horwell, SM) with reduced chamber depth is used for this assay method. Usually the standard procedure for the determination of the sperm concentration includes that a portion of the ejaculate is diluted with a fixative. The specimen is thoroughly mixed and a drop is transfered into a haemocytometer. The immobilized spermatozoa are allowed to sediment prior to counting. Analysis of sperm motility is done in parallel by estimating the proportion of motile sperm in an unfixed sample. Additionally, the progressive movement is arbitrarily scored on a scale from 1 to 4.

It is now feasible by computer-assisted image analysis to assay sperm concentration as well as sperm motion in a more precise way.

The objective of our study was to analyze how computer-assisted semen analysis is effected by the type and handling of a counting chamber because the optical and technical properties of the counting chamber can highly influence the results. In particular, we wanted to find out why under some conditions spermatozoa concentrate in the center of the counting chamber within a 10- to 30-min interval.

Material and Method

Semen samples were obtained from males undergoing andrological examination as part of a routine workup. The ejaculate was allowed to liquefy for 20–30 min at 37 °C prior to analysis.

The Makler chamber was used as a counting chamber. Handling followed the instructions as recommended by the manufacturer. Briefly, the prewarmed chamber was filled with a maximum of 5 µl of a mixed semen sample. After placing the counting chamber on the heated microscope stage (37 °C), all routine measurements were carried out within 5 min. Microscopic observation used a 20 × negative phase-contrast objective.

The Makler chamber was positioned on the microscope stage so that a frame next to the center included portions of the engraved network. The network line acted as a marker that the same frame was analyzed at 5-min intervals during the 30-min observation period. At each time interval this frame was analyzed three times and the average number of spermatozoa per frame was calculated. Next to the total number of spermatozoa per frame also the number of immotile and progressive motile sperm (forward movement with a speed $\geq$ 30 µm/s) were recorded.

All measurements were done with the SM motion analyzer (Strömberg-Mika, Bad Feilnbach, FRG). This system is the most advanced image analysis system for computer-assisted semen analysis. New hardware elements like a high resolution digitizer card and newly developed algorithm for functions like automatic grey level setting, object recognition and differentiation of immotile spermatozoa by tail detection as well as correct track analysis of crossing spermatozoa improved the analysis data markedly.

Results and Discussion

For better comparison the average number of spermatozoa per frame and not the sperm concentration in millions/ml were analyzed. Figures 1–3 display the factor time on the x-axis and the average number of spermatozoa per frame on the y-axis. In figure 1, the increase of the number of spermatozoa with time is shown for the total number of spermatozoa, progressive motile sperm, and immotile sperm. For each group of spermatozoa a regression line was calculated. All three sperm fractions show an increase with time. In particular, the increase of the number of immotile sperm argues against the hypothesis that active swimming of spermatozoa towards the center is responsible for an increase. Also, data in figure 2 indicate that active swimming is indeed not the cause for an increase of the number of spermatozoa in the center of the Makler chamber. The total number of sperm per frame nearly doubled within 30 min despite the fact that only 1 % of the spermatozoa were motile. A second series of measurements from the same semen sample showed no increase in the number of sperm over time. In fact, the measurements at 0 and 5 min gave identical results which underlined the accuracy of the SM motion analyzer.

Why do we observe in some cases an increase in the number of sperm when the time interval between filling of the counting chamber and analysis increases. If one reexamines the handling of the Makler chamber one realizes that very often the top of the counting chamber loosens its seat on the glass pillars. Under those circumstances suction forces (e.g. opposite to those during loading of the Makler chamber) contribute to a flow of fluid and cells toward the center.

Experimentally, these conditions can be tested by analyzing semen samples in a Makler chamber with a loose compared to a firmly seated top. An example of such measurements is show in figure 3. A semen sample from the same male was analyzed twice. As the data indicate, with a loose top the number of spermatozoa per frame increases with time compared to when the top rests firmly on the glass pillars of the Makler chamber. Again, at time 0–15 min, the two experiments led to similar results, underlining the reproducibility of the computer-assisted semen analysis with the SM motion analyzer.

The results can be summarized that analyzing sperm concentration with the Makler chamber is subjected to error if the top of the chamber does not rest firmly on the glass pillars and if the time interval between filling of the chamber and analysis is more then approximately 10 min. A loose chamber top effectively increases the depth of the counting chamber leading to the presence and object recognition of more spermatozoa per frame. Therefore, calculated sperm densities (in millions/ml) will reach a far higher number than actually present in the semen sample.

Fig. 1. Increase of the number of spermatozoa per frame as measured by computer-assisted semen analysis. The same frame of a Makler chamber was analyzed at 5-min intervals. The results are shown for the total number of spermatozoa (△), progressive motile sperm (○) and immotile sperm (●).

Fig. 2. Total number of spermatozoa per frame as measured by computer-assisted analysis. The top graph (▲) shows the results of the first filling of a Makler chamber with a semen sample containing less than 1 % motile spermatozoa. The bottom graph (□) displays the results obtained with the same semen samples during a second series of measurements.

Fig. 3. Total number of spermatozoa per frame as measured by computer-assisted analysis. The top graph (▲) shows the data points from measuring a Makler chamber with a loose top. The bottom graph (○) displays the results analyzing the same semen sample in a Makler chamber where the top is correctly and firmly seated on the glass pillars.

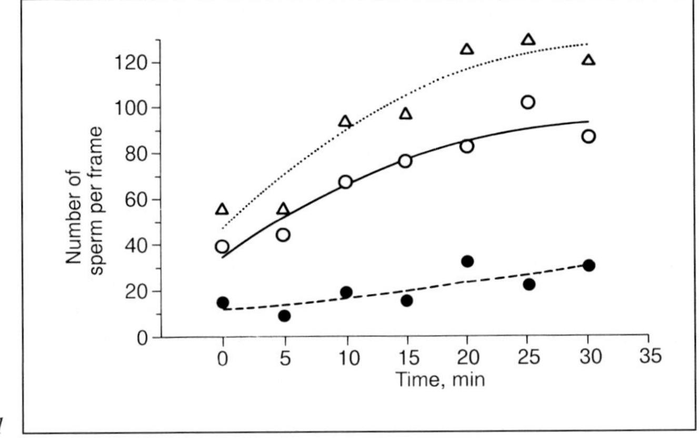

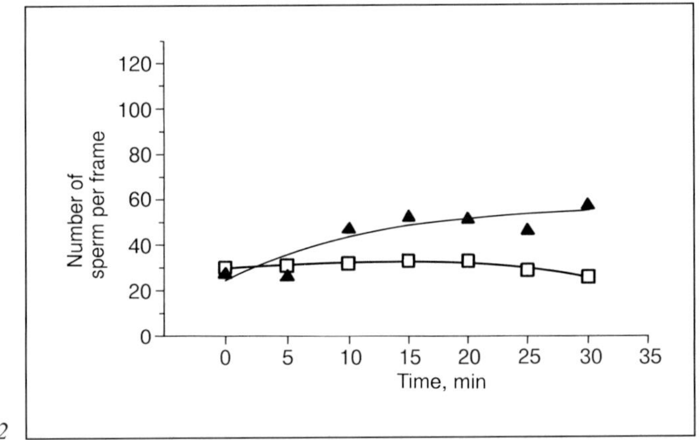

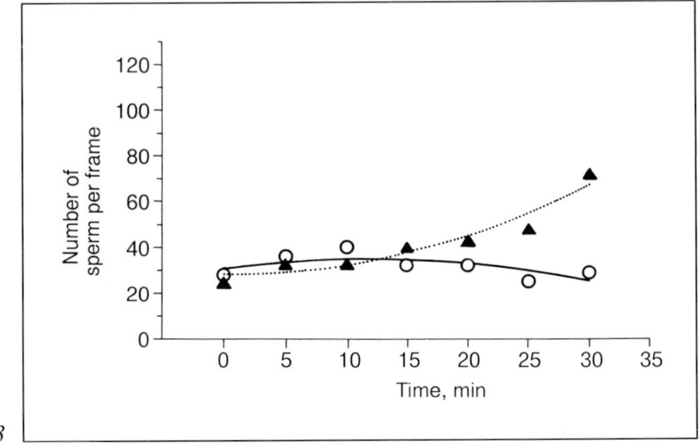

Table 1. Comparison of the area of contact between top and bottom in various counting chambers

Horwell	439 mm^2
Haemocytometer	80 mm^2
Makler	20 mm^2

The instability of the Makler chamber is probably due to the relatively small area where the bottom and top of the chamber are in contact. Table 1 shows for comparison the surface area, where the bottoms and tops of various chambers are in contact with each other. It is obvious that the larger that surface area is, the more stable during handling a counting chamber will be. It is at least our observation that an increase of the number of spermatozoa over time is far less often observed with counting chambers which follow the classical haemocytometer design. Nevertheless, the Makler chamber has its undisputed ease of use providing it is handled the proper way.

Dr. U. Schneider, Institute for Reproductive Medicine, Magdeburger Strasse 13, D-W–3257 Springe 1 (FRG)

Colpi GM, Pozza D (eds): Diagnosing Male Infertility.
Prog Reprod Biol Med. Basel, Karger, 1992, vol 15, pp 119–129

Male Distal Genital Tract Ultrasonography in Excretory Infertility

G.M. Colpi[a], *L. Negri*[a], *M.E. Mariani*[a], *K. Aydos*[a], *C. Del Favero*[b]

[a] Second Surgery Department, Andrology and Infertility Center, and
[b] Radiology Department, 'Valduce' Hospital, Como, Italy

During the last few years, prostate ultrasonography has been used in urology for diagnosing and monitoring of tumoral and inflammatory pathologies. In 1982, Jimenez-Cruz et al. [18] suggested the use of transabdominal seminal vesicle and prostate ultrasonography in some forms of excretory infertility. Colpi et al. [4–6] associated some pathological ultrasonographic images with specific clinical and seminal data, with the sperm count in urine after seminal tract washout, and with the related vasovesiculographic images, thus showing the diagnostic usefulness and reliability of ultrasonography in some forms of infertility induced by a defective (either absent or not complete) voiding of the distal seminal tract during ejaculation, in connection with anatomic or functional [17] pathologies of the same.

The use of high-resolution transrectal probes has further improved the diagnostic opportunities provided by this investigation [23, 38, 40] and has also allowed to achieve – in some selected cases of ampullovesicular voiding defects due to anomalies of the uroseminal carrefour – the restoration of male fertility, applying treatment based on ultrasonically guided transrectal puncture [7].

Although, at the present time, scientific literature on infertility caused by distal seminal tract (sub)obstructions is increasing [13, 16, 26, 29–31, 34–36, 41], very few data on the actual prevalence of this obstructive pathology in the infertile male population are available. The aim of this paper is to define precisely the correlation between the ultrasonographic and the vasovesiculographic images in the anatomic and in the functional voiding defects of the uroseminal carrefour, and to evaluate the prevalence of these pathologies in a group of patients in their fertile age.

Materials and Methods

The ultrasonographic investigation was carried out using a Brüel & Kjaer console type 1846, supplied with 4–7 MHz transrectal radial probe type 1850 and 7 MHz rectal sector transducer type 8537. All patients had a sexual abstinence of less than 48 h and full bladder; the examination was performed with the patient lying on his left side. In case of any pathological ultrasonographic evidence, a control was made 1 month later and in case of any further confirmation, the patient was subjected to vasovesiculography (in selected cases).

We evaluated the prevalence of ultrasonographic anomalies of the uroseminal carrefour in two different populations:

Group 1: 619 subjects, aged between 23 and 43 years, who underwent transrectal seminal vesicle and prostate ultrasonography (January–December 1989) in connection with different andrological troubles (infertility, prostatitis symptoms, testicular pain, micturition disorders with normal-size prostate at rectal touch, ejaculatory disturbances, hematospermia, premature ejaculation and other ejaculatory disorders).

Group 2: 242 male subjects of infertile couples at random, who, during the first 5 months of 1990, had completed a diagnostic iter inclusive of two semen analyses, MAR test IgG, Kremer test in serum AB Rh+ [20], basal hormonal (and, in selected cases, dynamic) screening, Doppler of spermatic vasa, Meares-Stamey test [22] or cytological examination of the expressed prostatic secretion. The patients affected by seminal tract infection or by prostatic inflammation were excluded from the group in order to prevent any doubtful interpretation of the nature of possible anomalies of the seminal vesicle echostructure. These patients were further subdivided into two subgroups, following the results of the semen analyses performed after 3–5 days of sexual abstinence: *Subgroup A:* normozoospermia or slight oligoasthenoteratozoospermia (OAT) and semen volume ≥ 2.5 ml; *Subgroup B:* medium or severe OAT, and/or semen volume < 2.5 ml.

Results

The anomalies of the uroseminal tract which we found most frequently by transrectal ultrasonography and further confirmed by vasovesiculography and, whereas necessary, by additional investigations (urethrocystography, urethrocystoscopy, NMR) were:

Utricular Cyst

It appears as an intraprostatic roundish anechoic area, located in a median position, 2–5 mm from the gland back border. According to our

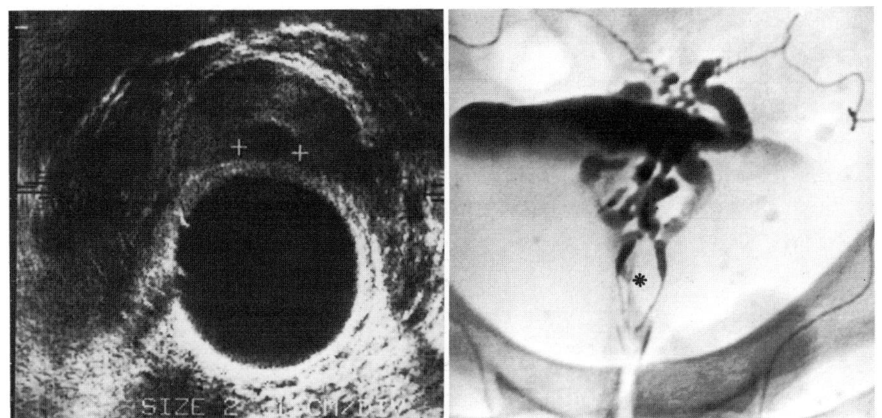

Fig. 1. Transrectal seminal vesicle and prostate ultrasonography (7 MHz Brüel & Kjaer radial probe). Medium-sized utricular cyst (+ + = 14 mm) within the prostatic parenchyma.

Fig. 2. Same case: digital vasovesiculography. Both ejaculatory ducts appear laterally displaced and their distal segment compressed by the utricular cyst (*), which does not communicate with the seminal tracts. Secondary infertility: gradual decrease of semen volume during the latest years; severe OAT; reduced vesicular biochemical markers.

experience, small-sized cysts (diameter < 4 mm) do not usually involve any consequence on the seminal tract voiding and therefore the ejaculate volume; while cysts with a diameter exceeding 10 mm always end by provoking a compression on the ejaculatory ducts, with the consequent vesicular and/or ampullar dilatation fig. 1, 2). Cysts of intermediate diameter are susceptible of causing partial or total voiding disturbance of the ampullovesicular tract. Ultrasonography does not usually allow to detect a possible ectopic outlet of one or both ejaculatory ducts into the cysts.

Functional Voiding Disturbance of the Ampullovesicular Tract

Seminal vesicles show a front-to-back transversal diameter > 13 mm, with internal hypoechoic polycyclic areas, even after a sexual abstinence lasting less than 48 h and in the absence of inflammatory processes – either past or in progress – which may be evidenced anamnestically and by Meares-Stamey test; the prostate does not show any echostructural changes resulting from a previous or presently existing inflammation (fig. 3, 4).

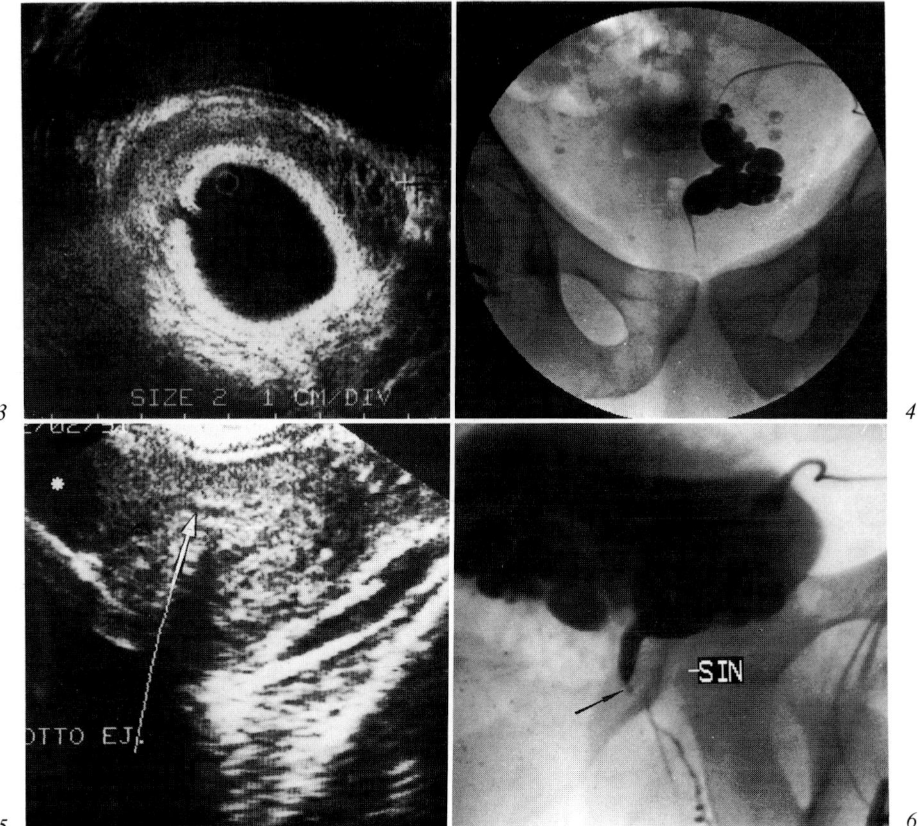

Fig. 3. Transrectal ultrasonography: the left seminal vesicle appears dilated (transversal diameter: 17 mm), with honeycomb anechoic areas.

Fig. 4. Left vasovesiculography: the ampullovesicular tract appears multiloculated and dilated with a difficult passage of the iodine medium through a spastic ejaculatory duct. The patient showed low semen volume and severe OAT; the right testis, previously cryptorchid, was atrophic; spermioculture and Meares-Stamey test resulted negative; no clinical data about previous inflammatory disorders affecting the urinary tract.

Fig. 5. Transrectal ultrasonography: a small hypoechoic strip with hyperechoic borders running along the left ejaculatory duct (arrow) can be seen; the ampullovesicular tract appears dilated and hypoechoic (∗); evident signs of a previous prostatic inflammation are present.

Fig. 6. Same case: vasovesiculography. The seminal tracts appear dilated and, although anatomically pervious, oppose resistance to the injected iodine medium (causing a retrograde epididymography). A substenosis can be observed in the distal portion of the left ejaculatory duct (arrow), involving a dilatation of the most cranial tract. Severe OAT with low semen volume; postejaculatory perineal pain; sporadic hematospermia in the past.

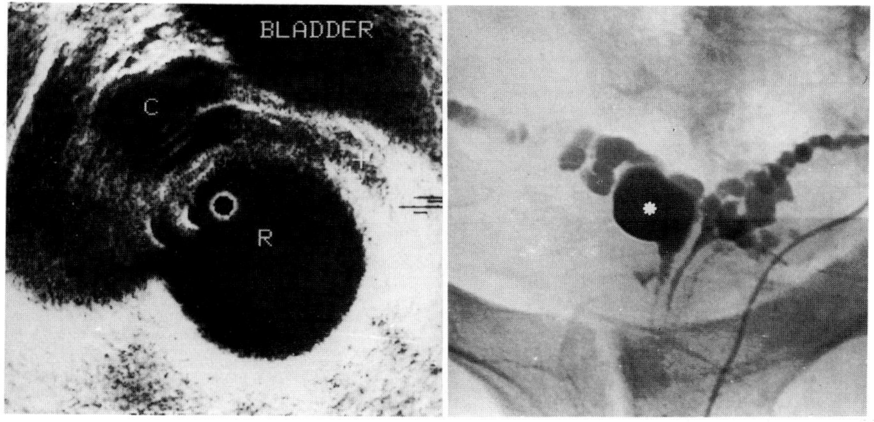

7

8

Fig. 7. Transrectal ultrasonography: anechoic multiloculated image consistent with a right vesicular cyst (C = cyst; R = rectum). Right epididymis partial agenesis; right kidney-ureter-hemitrigone agenesis; wide fluctuations of the ejaculatory volume presumably due to extrinsic compression exerted by the cyst on the left seminal tract.

Fig. 8. Bilateral vasovesiculography: the right deferential ampulla is normal, while the seminal vesicle (∗) appears cystic. The left seminal tract looks normal although partially displaced by the vesicular cyst. The shape of both ejaculatory ducts is normal. Severe hypospermia and OAT in a subject submitted to surgery for imperforate anus: transrectal ultrasonography could not be performed.

Ectasia or Pseudocystic Dysplasia of the Ejaculatory Duct

In radial scans, it appears as a roundish anechoic area, with a diameter of 3–8 mm, located in an intraprostatic paramedian position; in sagittal scans as a hypoanechoic dilatation of the ejaculatory duct image with a blind or 'rat-tail'-like bottom. Some postinflammatory alterations may be found, sometimes at the distal portion of the dilated ejaculatory duct. In this case, a concomitant dilatation of the seminal duct upstream is usually observed (fig. 5, 6).

Vesicular Cyst

It is a rare finding, appearing as an anechoic retrovesically developing sac-like area, partially involving the prostate even in the most caudal scans. The contralateral seminal duct is often displaced and enlarged owing to the extrinsic compression exerted by the cyst. The concomitant agenesis of the homolateral urinary tract (kidney, ureter, hemitrigone) is common [2, 11, 14, 15, 19, 32] (fig. 7, 8).

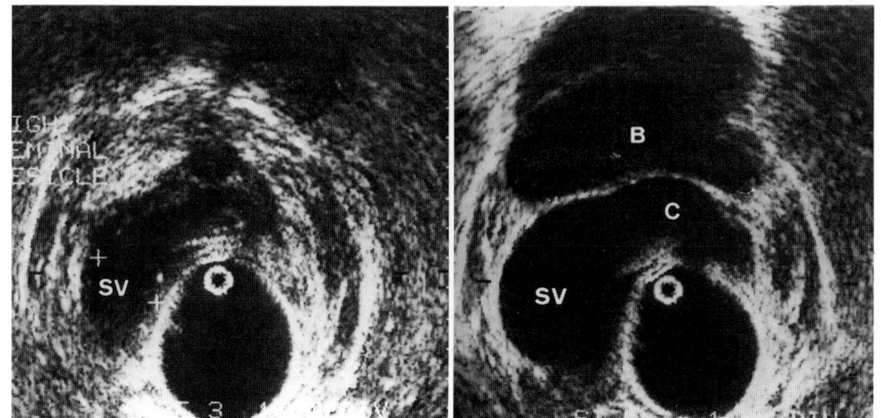

a   b

Fig. 9. a, b Transrectal ultrasonography: the right seminal vesicle in the caudal scans appears largely dilated while it seems to join a median supraprostatic retrovesical cyst in most cranial scans (SV = seminal vesicle; C = cyst; B = bladder).

Table 1. Anomalies of the uroseminal carrefour found by transrectal ultrasonography in 619 patients (age 18–43 years) affected by various andrological pathologies (see text)

Anomaly	Group 1 patients
Median intraprostatic anechoic area (diameter < 4 mm)	14 (2.2%)
Median intraprostatic anechoic area (diameter > 4 mm)	21 (3.3%)
Median or paramedian intraprostatic anechoic area plus intravesicular honeycomb anechoic areas	19 (3.0%)
Only intravesicular anechoic areas suggestive of glandular voiding dysfunction	36 (5.8%)
Total	90 (14.3%)

(21 (3.3%), 19 (3.0%), 36 (5.8%) bracketed together = 12.1%)

Müllerian Cyst

One report only, out of our filed cases resulting as a medium to large-sized hypoanechoic image, in a median retrovesical position, with the long peduncule originating from the prostate basis (fig. 9a, b, 10a–c).

The prevalence of ultrasonographic findings consistent with anatomic or functional anomalies of the distal seminal tract in group 1 is reported in table 1; the values referred to group 2, as well as to subpopulations 2A and 2B, respectively, are reported in table 2.

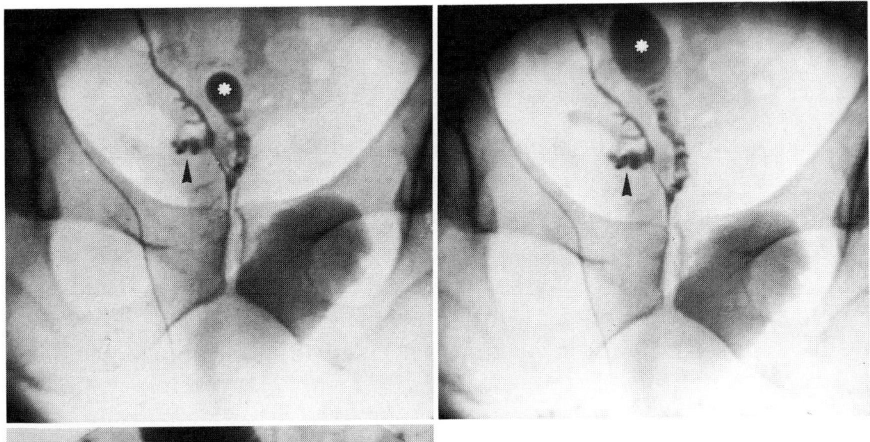
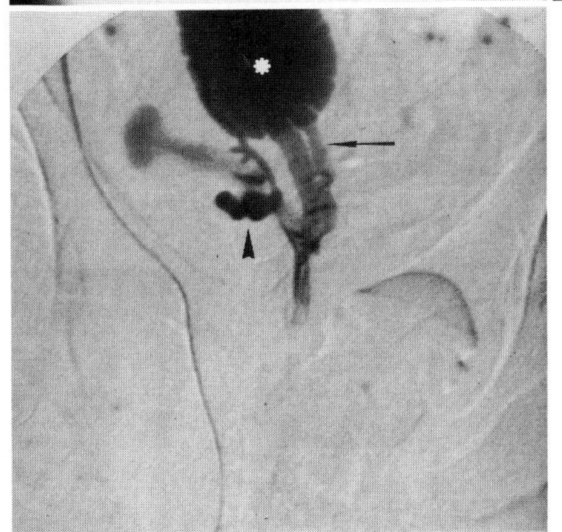

a

b

c

Fig. 10. a–c Right vasovesiculography: injection of iodine medium shows filling up of a large median cyst (*) located cranially to the prostate, communicating with the seminal tract through two long peduncles (arrow) which end up near the confluence of the seminal vesicle and the ejaculatory duct. The seminal vesicle (arrowhead), which ultrasonographically looked cyst-like, appears partially contrasted, presumptively owing to the stasis of seminal material in its inside. Secondary infertility: recurrent hematospermia; low semen volume with OAT and necrozoospermia.

Table 2. Anomalies of the uroseminal carrefour in 242 males of infertile couples (group 2), subdivided into subgroup 2A (normozoospermia or slight OAT and semen volume > 2.5 ml) and subgroup 2B (medium or severe OAT and semen volume < 2.5 ml)

Anomaly	Group 2 (n = 242)	Subgroup 2A (n = 64)	Subgroup 2B (n = 178)	
Median intraprostatic anechoic area (diameter < 4 mm)	5 (2.0%)	1 (1.5%)	4 (2.2%)	
Median intraprostatic anechoic area (diameter > 4 mm)	9 (3.7%)	3 (4.6%)	6 (3.3%)	
Median or paramedian intraprostatic anechoic area plus intravesicular honeycomb anechoic areas	18 (7.4%)	0* (0.0%)	18* (10.1%)	*$p < 0.05$
Only intravesicular anechoic areas suggestive of glandular voiding dysfunction	19 (7.8%)	1* (1.5%)	18* (10.1%)	*$p < 0.05$

Discussion

The scientific literature on the anomalies of the uroseminal carrefour of the last 30 years is mainly concerned with case reports about the most severe malformations giving rise, above all, to urological symptoms [1, 3, 8–10, 12, 21, 24, 25, 27, 28, 33, 37, 39, 42, 43]. Only in recent times, some papers were published where the presence of relevant Wolffian or Müllerian anomalies was correlated with severe male infertility [16, 26, 29–31, 34–36, 41]. In our opinion, the apparent rarity of these anomalies should be attributed to the three following factors:

(1) Most of the subjects affected by minor Wolffian or Müllerian anomalies do not complain about any symptoms.

(2) Most anomalies have so far escaped the ultrasonographists' attention, both for the poor definition of low-frequency transabdominal probes and for the anawareness of their existence (utricular cysts or ductal ectasia are often incorrectly labelled as 'retention cysts' or as 'anechoic area of doubtful interpretation'; in the same way, the anechoic polycyclic intravesicular and/or ampullar images – typical features of voiding disturbances – are often reported as acute, subacute or chronic vesiculitis, despite the fact that the clinical, cultural and cytological data related to the seminal secretions are absolutely negative).

(3) The little inclination of the nonurological andrologists to consider these pathologies.

Out of 100 subjects who underwent radical prostatectomy, being affected by malignant tumors, in whom the ultrasonographic and the histological data were correlated, Villers et al. [40] found utricular cysts (diameter > 4 mm) in 2% of the patients, ectasia of the ejaculatory ducts in 2%, dilatation of seminal vesicles in 2%, dilatation of ejaculatory ducts and of seminal vesicles in 5% (total percentage: 11%).

In our 619 andrological patients, we found a prevalence of ultrasonographic pictures consistent with anomalies of the uroseminal carrefour of 14.3% (12.1% if we exclude utricular cysts with a diameter $\leq$ 4 mm) (table 1), superimposable upon the results reported by Villers et al. [40] in 100 subjects presumably older than our patients.

The prevalence of utricular cysts found by ultrasonography and confirmed by vasovesiculography does not differ significantly in the two infertile subpopulations (4/64 vs. 10/178); conversely, the presence of prostatic median or paramedian cysts, associated with vesicular ectasia, or of only ampullovesicular dilatation with honeycomb anechoic areas appears significantly larger in group 2B (36/178) than in group 2A (1/64) (p < 0.05). In conclusion, images of seminal vesicle dilatation with honeycomb anechoic areas, just after recent ejaculation, at transrectal ultrasonography, must lead us to suspect a voiding disorder of the seminal tract and suggest other investigations (semen analysis after prolonged sexual abstinence [7], or after transrectal voiding puncture of a concomitant median anechoic area (utricular cyst?) [7], or digital vasovesiculography associated with the sperm count in the urine after seminal tract washout [6] and/or a postejaculatory vesiculographic image).

References

1   Aragona, F.; Spinelli, C.; Campatelli, A.; D'Elia, F.: Urinary retention due to Müllerian duct cyst: role of ultrasonically fine needle aspiration in diagnosis and treatment. J. Urol. *134:* 364–366 (1985).
2   Beeby, D.I.: Seminal vesicle cyst associated with ipsilateral renal agenesis: case report and review of literature. J. Urol. *112:* 120–122 (1974).
3   Brooks, T.R., Jr.: Cyst of the ejaculatory duct, case report. J. Urol. *101:* 881–883 (1969).
4   Colpi, G.M.; Balerna, M.; Ballerini, G.; Langé, A.; Beretta, G.; Campana, A.; Zanollo, A.: A new approach to diagnosis of seminal duct pathology: multiple non-invasive investigations; in Book of Abstracts of the XIth World Congress Fertil. Steril., Dublin 1983, p. 248.
5   Colpi, G.M.; Ballerini, G.; Zanollo, A.: Ultrasonography of seminal vesicles in infertility; in Hubimont, P.O. (ed.): Progress in Reproductive Biology and Medicine: Seminal Vesicles and Fertility, pp. 124–142 (Karger, Basel 1985).
6   Colpi, G.M.; Casella, F.; Zanollo, A.; Ballerini, G.; Balerna, M.; Campana, A.; Langé, A.: Functional voiding disturbances of the ampullo-vesicular seminal tract: a cause of male infertility. Acta Eur. Fertil. *18:* 165–179 (1987).
7   Colpi, G.M.; Negri, L.; Mariani, M.; Balerna, M.: Semen anomalies due to voiding defects of ampullo-vesicular tract. Andrologia *22:* suppl. 1, pp. 206–218 (1990).
8   Currarino, G.: Large prostatic utricles and related structures, urogenital sinus and other forms of urethrovaginal confluence. J. Urol. *136:* 1270–1279 (1986).
9   Das, S.; Amar, A.D.: Ureteral ectopia into cystic seminal vesicle with ipsilateral renal dysgenesis and monorchia. J. Urol. *124:* 574–575 (1980).
10 Devine, C.J., Jr.; Gonzalez-Serva, L.; Stecker, J.F.; Devine, P.C.; Horton, C.E.: Utricular configuration in hypospadias and intersex. J. Urol. *123:* 407–411 (1980).
11 Eisenberg, D.; Luis-Jorge, J.C.; Himmelfarb, E.H.; Kulkarni, M.V.; Shaff, M.I.: Sonographic diagnosis of seminal vesicle cyst. J. Clin. Ultrasound *14:* 213–215 (1986).
12 Elder, J.S.; Mostwin, J.L.: Cyst of the ejaculatory duct/urogenital sinus. J. Urol. *132:* 768–771 (1984).

13 Farley, S.; Barnes, R.: Stenosis of ejaculatory ducts treated by endoscopic resection. J. Urol. *109:* 664–666 (1973).
14 Fuselier, H.A.; Peters, H.D.: Cyst of seminal vesicle with ipsilateral renal agenesis and ectopic ureter: case report. J. Urol. *116:* 833–835 (1976).
15 Hart, J.B.: A case of cyst of the seminal vesicle. J. Urol. *96:* 247–249 (1966).
16 Hassler, R.D.; Weber, C.H.: Oligospermia secondary to Müllerian duct cyst. Urology *9:* 386–388 (1978).
17 Ichijo, S.; Sigg, C.; Nagasawa, M.; Sirawa, Y.: Vasoseminal vesiculography before and after ejaculation. Urol. Int. *36:* 35–45 (1981).
18 Jimenez-Cruz, J.F.; Mayayo, T.; Lovaco, F.; Garcia, J.; Navio, S.; Romero-Aguirre, C.: Transabdominal sonography of seminal vesicles. J. Urol. *127:* 260–262 (1982).
19 Kimchi, D.; Wiesenfeld, A.: Cyst of seminal vesicle associated with ipsilateral renal agenesis: case report. J. Urol. *89:* 906–907 (1963).
20 Kremer, J.: In vitro sperm penetration in cervical mucus and AIH; in Emperaire, J.C.; Audebert, A.; Hafez, E.S.E. (eds.): Homologous Artificial Insemination (AIH), pp. 30–37 (Nijhoff, The Hague 1980).
21 Lucey, D.T.; McAninch, J.W.; Bunts, R.C.: Genital cysts of the male pelvis: case report of Müllerian and ejaculatory duct cyst in the same patient. J. Urol. *109:* 440–443 (1973).
22 Meares, E.M.; Stamey, T.A.: The diagnosis and management of bacterial prostatitis. Br. J. Urol. *44:* 175–179 (1972).
23 Menchini Fabris, G.F.; Carletti, C.; Paoli, R.; Sarteschi, M.: Testo-Atlante di Ecografia in Andrologia, pp. 205–265 (Rima, Firenze 1989).
24 Morgan, R.J.; Williams, D.I.; Pryor, J.P.: Müllerian duct renmants in the male. Br. J. Urol. *51:* 488–492 (1984).
25 Neustein, D.H.; Schutte, H.: Müllerian duct cyst: with report of a case. Br. J. Urol. *40:* 72–77 (1968).
26 Perez-Palaez, M.; Jeyendran, R.S.; Malvar, T.: A case report: obstruction at colliculus seminalis. Andrologia *20:* 447–449 (1988).
27 Peterson, E.N.: Association of transverse testicular ectopia and seminal vesicle cyst. J. Urol. *118:* 345–346 (1977).
28 Polse, S.; Edelbrock, H.: Prostatic utricular enlargement as a cause of vesical outlet obstruction in children. J. Urol. *100:* 329–332 (1968).
29 Pomerol, J.M.; Marina, S.: Urological aspects in the treatment of infertile men; in Mancini, R.E.; Martini, L.; (eds.): Male Fertility and Sterility, pp. 497–515 (Academic Press, London 1974).
30 Porch, P.P.: Aspermia owing to obstruction of distal ejaculatory duct and treatment by transurethral resection. J. Urol. *119:* 141–142 (1978).
31 Payne, S.R.; Pryor, J.P.; Parks, C.M.: Vasography, its indications and complications. Br. J. Urol. *57:* 215–217 (1985).
32 Reddy, N.Y.; Winter, C.C.: Cyst of the seminal vesicle: a case report and a review of the literature. J. Urol. *108:* 134–135 (1972).
33 Schurhke, T.D.; Kaplan, G.W.: Prostatic utricle cyst (Müllerian duct cyst). J. Urol. *119:* 765–767 (1978).
34 Sharlip, I.D.: Obstructive azoospermia or oligozoospermia due to Müllerian duct cyst. Fertil. Steril. *41:* 298–303 (1984).
35 Silber, S.J.: Ejaculatory duct obstruction. J. Urol. *124:* 294–297 (1980).

36 Stanley, K.E.; Jr.: Müllerian duct cyst variant: utricolocele and inclusion of ejaculatory ducts: report of a case. J. Urol. *102:* 233–235 (1969).
37 Szemes, G.C.; Rubin, J.R.: Squamous cell carcinoma in a Müllerian duct cyst. J. Urol. *100:* 40–43 (1968).
38 Tanahashi, Y.: Seminal vesicles; in Watanabe; Holmes; Holm; Goldberg (eds.): Diagnostic Ultrasound in Urology and Nephrology, pp. 141–147 (Igaku Shoin, Tokyo 1981).
39 Van Poppel, H.; Vereecken, R.; De Geeter, P.; Verduyn, H.: Hemospermia owing to utricular cyst: embryological summary and surgical review. J. Urol. *129:* 608–609 (1983).
40 Villers, A.; Terris, M.K.; McNeal, J.E.; Stamey, T.A.: Ultrasound anatomy of the prostate: the normal gland and anatomical variation. J. Urol. *143:* 732–738 (1990).
41 Weintraub, C.M.: Transurethral drainage of the seminal tract for obstruction, infection and infertility. Br. J. Urol. *52:* 220–225 (1980).
42 Weiman, P.J.; McClennan, B.L.: Computed tomography and ultrasonography in the evaluation of mesonephric duct anomalies. Urol. Radiol. *1:* 29–37 (1979).
43 Yamashita, T.; Watanabe, K.; Ogawa, A.: Retroprostatic midline cyst involving both ejaculatory ducts. Urol. Radiol. *7:* 178–179 (1985).

Dr. Giovanni M. Colpi, Corso Cristoforo Colombo 8, I–20144 Milano (Italy)

Colpi GM, Pozza D (eds): Diagnosing Male Infertility.
Prog Reprod Biol Med. Basel, Karger, 1992, vol 15, pp 130–135

Indications for Vasovesiculography and Testicular Biopsy

An Update

John P. Pryor

St. Peter's Hospital, London, UK

Indications for Testicular Biopsy or Exploration

The indications for testicular biopsy are given in table 1 and the majority are related to the investigation of the infertile male with poor quality semen. The advent of readily available radioimmunoassay techniques to estimate the level of follicle-stimulating hormone (FSH) has altered the indications for testicular biopsy. Those men with small testes (less than 4 cm long) and plasma FSH levels greater than twice normal have primary testicular failure and do not require testicular exploration [1, 2]. Those azoospermic men with normal-sized testes, or with small testes and without grossly elevated levels of plasma FSH, require testicular exploration in order to exclude an obstructive cause for the azoospermia. The epididymis is inspected for evidence of obstruction and vasography is performed. The outcome of the testicular exploration is shown in table 2.

The indications for testicular biopsy in oligozoospermic men are less well defined but the objective is to determine the cause for oligozoospermia in order to have a rational basis for its treatment. It is important to be able to exclude unilateral obstruction, whether it be to overcome the obstruction or treat an immunological cause of infertility [3]. In general terms, I do not think that it is worthwhile biopsying or exploring unless there is consistently less than 5 million sperms in the ejaculate. The correlation between testicular size, FSH level and histological assessment of spermatogenesis is less well defined in digozoospermia. Whilst it is true that some men with few sperm in the ejaculate have small testes and an elevated FSH level, others may have normal-sized testes and the oligozoospermia be due to numerous obstructive causes as shown in table 3.

Table 1. Indications for testicular biopsy

Mandatory	Azoospermia
	    if testes normal size
	    or FSH less than twice normal
	Severe oligozoospermia (< 5 million/ml)
	    as azoospermia

Incidental	Varicocele ligation Vasectomy reversal Torsion
Useful	Contralateral testis at tumour orchidectomy Orchidopexy

Table 2. Outcome of testicular exploration in 494 men

Outcome of exploration	Patients	
	n	%
Primary testicular failure	78	16
Impaired spermatogenesis	96	19
Normal spermatogenesis	320	65

Table 3. Causes of obstructive oligozoospermia

Site of obstruction	Causes
Intratesticular	congenital immunological infective hypercurvature
Epididymal	congenital dysplasia infective immunological
Vasal	infective adynamic
Ejaculatory duct	congenital anomalies infective

Specific histological changes have never been demonstrated in men with varicoceles and testicular biopsy in infertile men with normal numbers of sperm undergoing a high ligation of a varicocele is unnecessary. In those men with azoospermia or severe oligozoospermia undergoing varicocele ligation for infertility, the operation may be performed at the level of the inguinal canal and through such an incision a bilateral testicular exploration and vasography may be performed.

Testicular biopsy may be performed at the time of vasectomy reversal but is probably unnecessary as spermatogenesis is almost always preserved. Although the occurrence of carcinoma in situ is well recognized, the histological diagnosis remains difficult. The occurrence of bilateral germ cell tumours in perhaps 10% of patients makes it justifiable to biopsy the contralateral testis in those patients with a stage I tumour who may be managed by surveillance. Elective testicular biopsy during chemotherapy for a childhood leukaemia is no longer considered to be beneficial [4].

Because of the well-known association of testicular germ cell tumours and maldescent it is useful to biopsy the testis. Carcinoma in situ was found in 29% of intra-abdominal testes, 7.5% of inguinal testes and none of the scrotal testes when biopsied after puberty [5].

Testicular exploration is usually performed through a single midline scrotal incision and the tunica vaginalis of each testis incised and the testis allowed to prolapse into the wound. The epididymis is next inspected in an attempt to determine whether it is obstructed. High obstruction (vasal or ejaculatory duct) may not cause much distension of the epididymis – the thick muscular wall of the vas deferens preventing it – and may be overlooked. Clinical inspection of the epididymis gave an excellent assessment of obstruction in 258 men [2]. It should be remembered that clear fluid may be present in the dilated epididymal tubules in the presence of multiple obstructions or occasionally as an artefact (anaesthetic-induced?).

Vasography is conveniently performed at the time of testicular exploration by placing a sling around the vas (00 suture) in order to angle the thick wall and facilitate puncture with a 25-gauge needle mounted on a 10-ml syringe. The needle may be introduced through the vas in the direction of its lumen. A water-soluble contrast medium is used (I still use 45% hypaque) and 2–3 ml is injected towards each seminal vesicle. It is not satisfactory to rely upon the free injection of saline as congenital abnormalities would be missed. It is useful to open the pelvis by angling the radiological beam by 10° to allow the ejaculatory ducts to be visualized.

Use of an image intensifier speeds up the procedure but it is still useful to have good quality films available. The testes may be biopsied whilst awaiting the development of these films. It has not been found useful to inject towards the epididymis as obstruction at this level is better determined by visual inspection.

A small (2–3 mm) incision is made in the tunica albuginea and the testicular tissue is prolapsed through the incision by gently squeezing the testis. The tissue is excised with a sharp scissors and placed directly into a preservative (such as Bouin's) which does not cause shrinkage and distortion of the seminiferous tubules. It is important not to traumatize the biopsy by squeezing it with forceps.

It is beyond the remit of this article to discuss the techniques of making the histological assessment, but some attempt should be made to quantify the changes. Our laboratory uses the Johnsen technique [6] but there are some advantages in using semi-thin sections of the testis.

Safety and Outcome of Testicular Exploration and Vasography

In a review of 494 testicular biopsies from azoospermic men, primary testicular failure occurred in 78 (16%), impaired spermatogenesis in 19% and normal spermatogenesis in 320 (65%). These biopsies were taken before the significance of small testes and a high FSH level was well known. The level of obstruction in those men with normal spermatogenesis is shown in table 4. In a more recent study, ejaculatory duct obstruction

Table 4. Causes of obstruction in 321 azoospermic men with normal spermatogenesis

Cause of obstruction	Patients	
	n	%
Intratesticular obstruction	47	15
Epididymal obstruction	163	51
Bilateral vasa aplasia	58	18
Unilateral vasa aplasia and other obstruction	11	3
Vasal obstruction	39	12
Ejaculatory duct obstruction	3	1

Table 5. Causes of ejaculatory duct obstruction in 69 patients

Cause of obstruction	Patients
Congenital midline Müllerian abnormality	10
Congenital lateral (Wolffian) abnormality	24
Traumatic	5
Infective	16
Neoplastic	2
Adynamic	6
Other	6

Table 6. Incidence of abnormality at vasography in 482 patients

Indication	Number	Abnormal	
		n	%
Azoospermia	263	26	10
Oligozoospermia	177	7	4
Miscellaneous	42	22	52
Total	482	55	12

was found in 80 patients and the cause of this is shown in table 5. It is important to remember that ejaculatory duct obstruction was associated with epididymal obstruction in 10% of patients.

The safety of vasography has been the subject of much debate but in a series of 154 infertile men we found that the semen quality was unchanged in 96, improved in 29 and deteriorated in 29 [7]. No patient became azoospermic and in all instances the quality of the semen returned to the previous level. The incidence of abnormality at the time of vasography is shown in table 6. The high incidence of abnormalities in the group of 28 men with haemospermia was a result of patient selection. Vasography was only performed in those men in whom the haemospermia had persisted over a long period of time – usually more than 1 year.

Conclusions

The indications for testicular exploration with biopsy in infertile men are now well defined and are mandatory in the investigation of azoospermia or severe oligozoospermia. Vasography is a simple and safe procedure and is carried out at the time to exclude an obstructive problem. Any reconstructive procedure that is necessary may also be performed. The possibility of cacinoma in situ should always be remembered by the pathologist and the clinician and its detection and treatment may prevent the onset of a malignant tumour. Vasography has a small but important role in the investigation of persistent haemospermia or in men with ejaculatory dysfunction.

References

1   Pryor, J.P.; Pugh, R.C.B.; Cameron, K.M.; Newton, J.R.; Collins, W.P.: Plasma gonadotrophic hormones, testicular biopsy and seminal analysis in the men of infertile marriages. Br. J. Urol. *48:* 709–717 (1976).

2   Pryor, J.P.; Cameron, K.M.; Collins, W.P.; Hirsh, A.V.; Mahony, J.D.H.; Pugh, R.C.B.; Fitzpatrick, J.M.: Indications for testicular biopsy or exploration in azoospermia. Br. J. Urol. *50:* 591–594 (1978).

3   Hendry, W.F.: Clinical significance of unilateral testicular obstruction in subfertile men. Br. J. Urol. *58:* 709–714 (1986).

4   Pui, L.H.; Bowman, W.P.; Abromowitch, M.; Dahl, G.V.; Ochs, J.; Rivera, G.: Elective testicular biopsy during chemotherapy for childhood leukaemia is of no clinical value. Lancet *ii:* 410–412 (1985).

5   Pryor, J.P. ; Cameron, K.M.; Chilton, C.P.; Ford, T.F.; Parkinson, M.C.; Sinokrot, J.; Westwood, C.A.: Carcinoma in situ in testicular biopsies form men presenting with infertility. Br. J. Urol. *55:* 780–784 (1983).

6   Johnsen, S.G.: Testicular biolpsy score count. Hormones *1:* 2–25 (1970).

7   Payne, S.R.; Pryor, J.P.; Parks, C.M.: Vasography, its indications and complications. Br. J. Urol. *57:* 215–217 (1985).

Dr. John P. Pryor, St. Peter's Hospital, Henrietta Street,
London WC2E 8NE (UK)

Colpi GM, Pozza D (eds): Diagnosing Male Infertility.
Prog Reprod Biol Med. Basel, Karger, 1992, vol 15, pp 136–141

Unexpected Findings of Scrotal Exploration during Varicocele Surgery in Infertile Patients

D. Pozza[a], *L. Marchionni*[a], *P. Ossanna*[a], *A. Gregori*[a], *M.E. Hallgass*[a], *G. Spera*[b]

[a] Centro di Andrologia e di Chirurgia Andrologica, and [b] V Clinica Medica, University of Rome, Italy

Varicocele is commonly acknowledged as one of the major causes of oligozoospermia. The etiology of the spermatogenetic damage is not well understood even if many pathogenetic mechanisms such as high temperature, hypoxia, reflux of adrenergic substances, modifications of the epididymal and testicular circulation have been postulated. None of them has been clearly established [1]. The results of the surgical treatment seem to confirm the role played by varicocele. Actually, varicocelectomy is accompanied by clear improvement of the seminal parameters in 30–65% of the cases [2].

We do not exactly know why 35–70% of patients do not recover after surgery. Possible causes of failure could be considered as related to an inaccurate etiological diagnosis of oligozoospermia, or to the presence of irreversible testicular lesions. In about 5–15% of the cases an inccorrect surgical treatment determines the persistence of varicocele [3, 4], but we can believe that in patients with oligozoospermia and varicocele more than one cause of oligozoospermia could be present.

Such causes are not always demonstrated by means of the common diagnostic procedure and could be shown only by histological examination. In our opinion, another aspect should not be underestimated: patients with varicocele could also be affected by epididymal or deferential diseases able to induce oligozoospermia. These elements could explain some cases of unsuccessful improvement of the seminal pattern after surgery.

The utility to perform a testicular biopsy before or during surgical operation for varicocele has been greatly questioned. Testicular biopsies have been performed by some authors before surgery to evaluate if the surgical treatment could restore a normal spermatogenesis. Results of such a diagnostic procedure were controversial [1, 5–7]. Other authors deny any prognostic value of the testicular biopsy in predicting the results of the surgical therapy. It is commonly thought that information deriving from biopsy is essentially poor. In the last years only a few surgeons perform a testicular biopsy on patients submitted to operation for varicocele.

We studied a group of 194 patients operated on for varicocele to verify the possible incidence of pathological lesions not strictly correlated with the varicocele itself. The existence of such lesions could explain some cases of insufficient improvement after surgery.

Material and Methods

From January 1987 to December 1989, 1,382 patients consulted our fertility clinic and were submitted to a diagnostic procedure considering anamnesis, clinical evaluation, seminal examinations on at least two or three samples, doppler flowmetry of the spermatic veins; FSH, LH, testosterone and prolactin values in basal conditions and after GNRH, when necessary. In many cases biochemical analyses of the seminal plasma (fructose, citric acid, acid phosphatase, GPC, proteins), bacteriological exams (often including chlamydia and mycoplasma research), immunological tests (MAR test, agglutinating Ab), testicular and prostatic sonography were also performed.

Varicocele was detected in 395 (28.5%) patients but only in 225 (16.2%) cases was it considered as the only evident cause of oligozoospermia. Surgical operation was suggested to all 225 infertile patients with varicocele. A group of 31 either refused operation or preferred to be operated on by other surgeons; 194 patients were submitted to surgery. The aim of our study was to evaluate the incidence of collateral pathological lesions and therefore to assess the usefulness of bilateral scrotal exploration (BSE) and bilateral testicular biopsy (BTB) during varicocele surgery. For this reason all patients operated on for varicocele were also submitted to BSE and BTB. A spermatic venography was performed before surgery in all patients to visualize the caliber, direction, number and collaterals of the spermatic veins [8].

The surgical operation consisted of high ligation and division of the spermatic veins.

BSE associated with BTB was performed through a scrotal incision in 123 patients and an infrapubic Kelami's incision in 71.

Results

Unexpected findings of BSE were 2 cases of testicular tumor (a dermoid cyst of the left testicle in one and a bilateral papillary cystoadenoma of the epididymis in another patient), 2 epididymal cysts which could determine tubular obstruction, 2 unilateral vas agenesis, 1 iatrogenic lesion of the vas and 2 obstructions, probably infectious in origin, of the vas at the cauda epididymis. The incidence of unexpected macroscopic findings was 4.6% (9 or 194 cases).

All testicular specimens were examined by the same pathologist. According to previous experience on testicular histology of varicocele patients [8], in 159 cases we observed what could be considered 'expected findings'. In 10 patients a quite normal histological picture was found. Most of our cases, 96, showed oligospermiogenesis, 10 hypospermatogenesis, in 20 cases associated lesions of oligospermiogenesis and hypospermatogenesis and in 23 cases maturative lesions at various levels were discovered.

In a group of 35 patients we found what we consider 'unexpected histological findings'. Seven patients showed unilateral Sertoli cell syndrome, 5 at the right testicle and 2 at the left one. Bilateral lymphocytic infiltration was found in 5 patients, epithelial lesions assimilable to carcinoma in situ in 3 and a clear-cut difference between left and right testicle was found in 20 patients. In most of these cases oligospermiogenesis was the main lesion on one side while various degree of maturative lesions were observed on the other side. The incidence of 'unexpected' microscopic findings was 18% (35 of 194).

In a few cases (5) macro- and microscopic findings were observed in the same patients. The overall incidence of unexpected findings was 20.1% (39 of 194 patients). If we consider our cases in a retrospective way we can say that some cases could be discovered before surgery with a more careful diagnostic procedure. One case of testicular tumor could be diagnosed by sonography. Two cases of vas agenesis could probably be revealed by clinical examination. Two cases of epididymal cysts could have been discovered by sonography. Actually, one of them had a scrotal sonography that did not reveal anything. In 3 cases of only Sertoli cell syndrome, we could suspect the histological diagnosis taking in account the small volume of the testicle. In 2 of these cases we discovered that the patients had history of retained testicle not referred at the first visit. So, in 8 cases a correct diagnosis would probably have been possible with a better examination.

Even if we exclude these 8 cases, in 31 cases (15.9%) the use of BSE and BTB allowed us to find other causes of oligozoospermia in patients operated on for varicocele and therefore to better understand the prognosis of the patients.

All operations were performed under general anaesthesia. Pre-operatory antibiotic prophylaxis with cefotetan, 1 g, was administered. One patient developed a scrotal haematoma that was treated 10 h after operation without any subsequent complication. Scrotal oedema was observed in a few patients. All patients were dismissed 2 or 3 days after surgery.

Conclusions

In the last 30 years a great amount of basic and clinical research has been spent to elucidate the relationships between varicocele and male infertility [1]. No clear-cut evidence has been found. Some authors reported their doubts about the real pathogenetic effects of varicocele upon male fertility [11, 12].

In their opinion patients with varicocele, even if not submitted to any treatment, would have the same, if not better, pregnancy rate than patients operated on. Such opinions are supported by the low incidence of positive results after treatment. Actually, the average improvement of the seminal parameters does not succeed 60%, while the pregnancy rate varies between 15 and 45%. If varicocele is responsible for oligozoospermia, why does treatment offer so limited results? We think that a proper selection of patients to be operated on represents the right answer to the problem [10, 13]. In our opinion not all patients with varicocele should be operated on. Azoospermic patients do not get any benefit from varicocele operation. Patients over 40 do not show good results after treatment. The presence of anti-sperm antibodies normally does not change after varicocele operation. Congenital or acquired lesions of the seminal pathways represent a not uncommon cause of oligozoospermia. In a group of 1,500 oligo- or azoospermic patients, submitted to scrotal exploration, we detected the presence of congenital anomalies in 7% [14]. The results of the present study reveal that 4.6% of patients affected by varicocele were also affected by congenital or acquired lesions of the seminal ducts that could sustain oligozoospermia. Such lesions were not detected by the routine diagnostic pre-operatory evaluation. Some of them could be treated with microsurgical technique. If not explored these patients would not get any improvement of the seminal picture even with a better spermatogenesis. It seems evident that a bilateral scrotal exploration during varicocele surgery could offer better chances of treating the oligozoospermic patients. Histological data are very important when oligozoospermia and varicocele are considered. In our experience it seems quite difficult to predict the results of surgery on the basis of the testicular histology. Sometimes pregnancies occur even with no significant modification of the sperm concentration, probably because a better epididymal transit of spermatozoa occurs. We have to underline that the 7 testicles with only Sertoli cell syndrome and the 20 with maturative blockade could not modify their spermatogenesis after varicocele operation. Another important aspect connected with scrotal exploration is that related to the oncological problem. In 2 patients we found macroscopic unexpected tumors of the testicles and in 3 cases carcinoma in situ-like lesions were detected at biopsy. If not explored or biopsied, these patients could have suffered subsequent troubles.

The routine performance of BSE and BTB in cases with oligozoospermia and varicocele is simple, has low morbidity and does not prolong hospitalization. We think that it would be nonsense to perform BSE and BTB before varicocele operation. If the patient is oligozoospermic, there is no evidence of infectious, immunological or hormonal disease and varicocele is detected, it seems worthwhile to submit the patient to high ligation and division of the spermatic veins in conjunction with a BSE and BTB. Obstructive diseases, incidentally detected, could be treated during the same operation.

For these reasons we think that BSE and BTB should be performed in any patient during varicocele surgery. They can improve our diagnostic and therapeutic ability in cases of oligozoospermia and varicocele.

References

1   Lewis RW, Harrison RM: Diagnosis and treatment of varicocele. Clin Obstet Gynecol 1982;25:501–523.
2   Glezerman M: A short historical review and comparative results of surgical treatment for varicocele; in Glezerman M, Jecht EW (eds): Varicocele and Male Infertility. II. Berlin, Springer, 1984, pp 87–93.
3   Pozza D, D'Ottavio G, Nanni E, Marchionni L, Zappavigna D: Il varicocele recidivo; in D'Ottavio G, Pozza D (eds): Aggiornamenti in Andrologia Chirurgica. Rome, Acta Medica, 1986, pp 285–295.
4   Jecht EW, Zeitler E: Varicocele and male infertility. Berlin, Springer, 1982.
5   Hadziselimovic F, Leibundgut B, Da Rugna D, Buser MW: The value of testicular biopsy in patients with varicocele. J Urol 1985;135:707–710.
6   Johnsen SG, Agger P: Quantitative evaluation of testicular biopsies before and after operation for varicocele. Fertil Steril 1977;29:58–63.
7   Dubin L, Hotchkiss RS: Testis biopsy in subfertile men with varicocele. Fertil Steril 1969;22:50–57.

8   Pozza D, D'Ottavio G, De Pace F, Pisanello M, Damia S, Alessi A, Zappavigna D: Varicocele ed infertilità maschile. Reperti flebografici preoperatori. Min Urol 1980; 32:23–30.
9   Spera G, Medolago-Albani A, Coia L, Morgia C, Gonnelli S, Ghilardi S: Histological, histochemical, and ultrastructural aspects of interstitial tissue from the contralateral testis in infertile men with monolateral varicocele. Arch Androl 1983;10: 73–78.
10   Pozza D, D'Ottavio G, Masci P, Coia L, Zappavigna D: Left varicocele at puberty. Urology 1983;22:271–274.
11   Nilsson S, Edvinsson A, Nilsson B: Improvement of semen and pregnancy rate after ligation and division of the internal spermatic vein: fact or fiction? Br J Urol 1979; 51:591–596.
12   Vermeulen A, Vandeweghe M: Improved fertility after varicocele correction: Fact or fiction? Fertil Steril 1984;42:249–256.
13   Pozza D, Marchionni L, Gregori A, Hallgass ME, Di Trapano C, Spera G: Unexplained infertility and varicocele; in Spera G, Gnessi L (eds): Unexplained Infertility: Basic and Clinical Aspects. New York, Raven Press, 1989, pp 149–153.
14   Pozza D, Marchionni L, D'Ottavio G: Congenital anomalies of the male seminal tract. Turk J Urol 1987;13:180.

Diego Pozza, Centro di Andrologia e di Chirurgia Andrologica,
via Flaminia Nuova 280, I–00136 Roma (Italy)

Colpi GM, Pozza D (eds): Diagnosing Male Infertility.
Prog Reprod Biol Med. Basel, Karger, 1992, vol 15, pp 142–147

Diagnosing Clinical and Subclinical Varicocele: Comparative Investigations by Doppler Sonography and Phlebography

H. Gall[a], *W. Bähren*[b]

Departments of [a] Dermatology and [b] Radiology, Federal Army Hospital
(Bundeswehrkrankenhaus), Ulm, FRG

Varicocele is caused by a retrograde blood flow running from the left renal vein (on the right side from inferior vena cava) into the testicular vein down to the pampiniform plexus [3]. Doppler sonography is an easy method to detect retrograde blood flow (reflux) in the testicular vein [5]. In contrast, phlebography represents an invasive technique to demonstrate reflux in the testicular vein [1, 3]. In this study, comparative investigations were done with Doppler sonography and phlebography to reveal reflux in the testicular vein. Thereby, the values of the two methods should be evaluated in diagnosing clinical and subclinical varicocele.

Materials and Methods

From 1982 to 1985, investigations by Doppler sonography and phlebography were performed in 330 patients with left-sided varicocele, thereby 74 varicoceles on the right side. Clinical varicocele was distributed into 3 sizes; subclinical varicocele was determined by Doppler sonography. Doppler sonography was done as part of the andrologic investigation and phlebography preceded the intended sclerotherapy of the testicular vein. For detection of retrograde blood flow in the testicular vein at the level of the spermatic cord, we used the directional Doppler flowmeter 762 by Kranzbühler. Percutaneous retrograde phlebography of the left renal vein was performed according to a modified technique from Zeitler et al. [7] to prove testicular vein insufficiency, on the right side in the inferior vena cava.

Table 1. Results of comparative investigations by Doppler sonography and phlebography

	Size of varicocele	Doppler sonography: reflux positive	Phlebography: testicular vein insufficiency
Left	subclinical	8	–
	small	40	30
	medium	156	150
	large	126	123
Right	subclinical	62	20
	small	12	8

Results

The results of the comparative investigations by Doppler sonography and phlebography are shown in table 1. In all patients with varicocele on the left and right side, Doppler examination detected reflux in the testicular vein. Phlebography of the left renal vein showed 5 types of testicular vein insufficiency (fig. 1), as described by Bähren et al. [1]. In subclinical varicocele phlebography revealed no testicular vein insufficiency on the left side. In small varicocele, phlebography confirmed reflux in 30 of 40 patients. Phlebography failed 6 times in medium varicocele and 3 times in large varicocele. On the right side, phlebography demonstrated testicular vein insufficiency in 20 of 62 cases of subclinical varicocele and in 8 of 12 cases in small varicocele.

Discussion

In subclinical varicocele there is a great discrepancy between Doppler sonography and phlebography, less in small varicocele. An explanation for these different findings is given in figure 2. While the patient is performing a Valsalva maneuver, you can detect reflux in the testicular vein at the level of the spermatic cord by means of Doppler sonography in the case of testicular vein insufficiency (a), but also in cases with a competent orifice valve (b, c). This is the reason for the different findings in subclin-

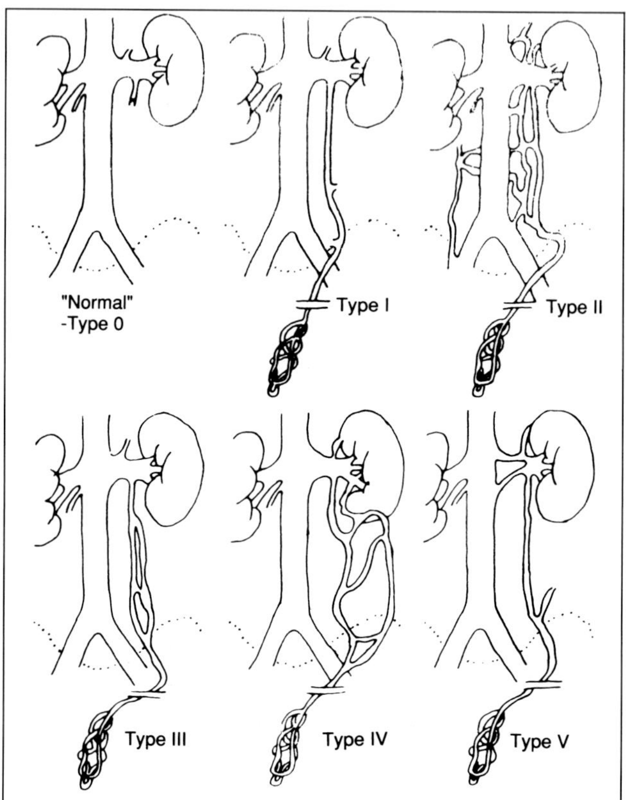

Fig. 1. Venographic types of left testicular vein insufficiency according to Bähren. Type 0 = Sufficient orifice valve; type I = only one insufficient trunk; type II = one insufficient trunk with collaterals to the lumbar plexus; type III = the trunk of the insufficient testicular vein is divided in two or more major branches, with cross-connections between the branches; this is called the 'ropeladder system'; type IV (a) insufficient trunk of the testicular vein and insufficient perirenal and renal segmental veins, and (b) the trunk has a competent valve bypassed by an insufficient collateral; type V, a venous ring composed of preaortic renal vein, retroaortic renal vein, and vena cava ('renal collar').

ical and small varicocele, where Doppler examination detects reflux in the testicular vein despite proven testicular vein competence by phlebography.

While subclinical and small varicocele is caused by Valsalva's maneuver, medium and large varicocele is caused by spontaneous reflux [4].

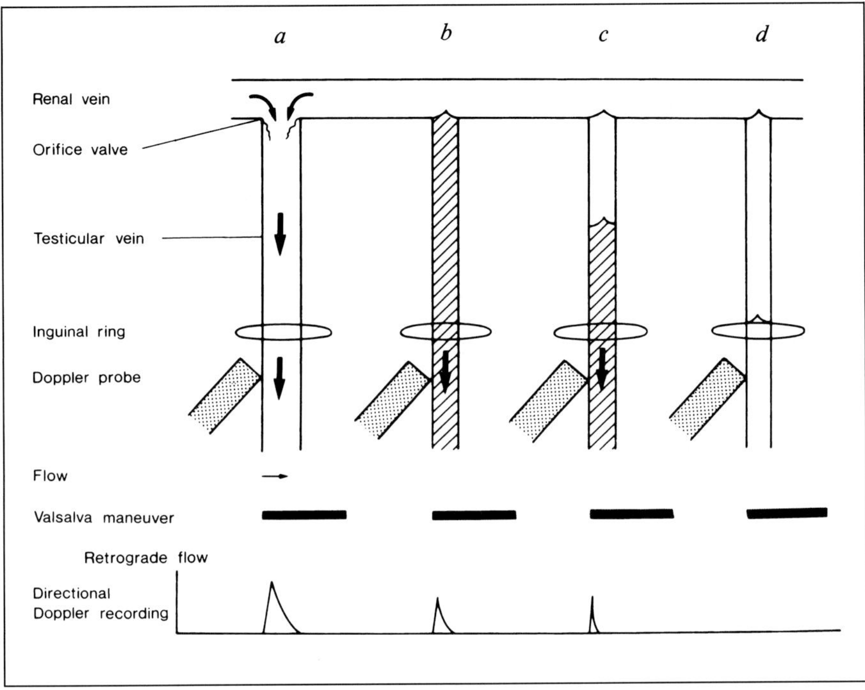

Fig. 2. Demonstration of reflux in the testicular vein at the level of spermatic cord by means of Doppler sonography. *a* Testicular vein insufficiency. *b, c* Without testicular vein insufficiency (competent orifice valve). *d* No reflux.

But spontaneous reflux is only possible in the case of testicular vein insufficiency. In medium and large varicocele phlebography could not demonstrate testicular vein insufficiency in 9 cases, whereas Doppler sonography detected reflux in the testicular vein at level of the spermatic cord. In this case the so-called 'distal reflux' is presumed [2]. However, phlebography could not reveal reflux running from the common iliac vein down to the pampiniform plexus via the deferential or cremasteric vein. The reflux demonstrated by Doppler sonography may be caused by type IVb according to Sigmund et al. [6] with a competent orifice valve bypassed by an insufficient collateral vein (fig. 3).

In conclusion, Doppler sonography is the most reliable method for detecting reflux in the testicular vein and phlebography to demonstrate testicular vein insufficiency in patients with clinical varicocele. There are

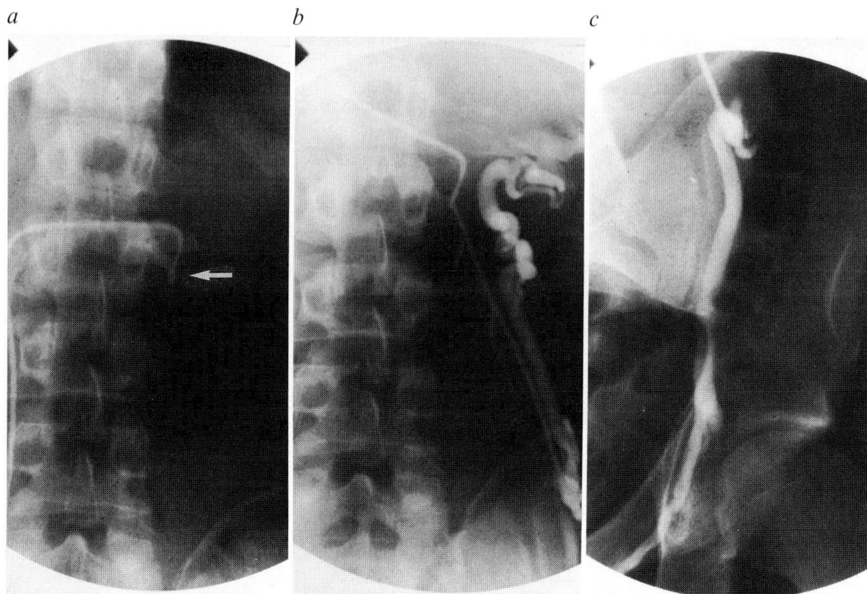

Fig. 3. Testicular vein insufficiency, type IVb. *a* Competent valve (arrow) in the proximal testicular vein near the orifice. *b* Selective probing of the testicular vein depicts insufficient collateral vein. *c* Contrast medium running down to pampiniform plexus.

two different meanings of subclinical varicocele determined by Doppler sonography (reflux in the testicular vein) and phlebography (testicular vein insufficiency with retrograde blood flow down to the pampiniform plexus). The so-called 'distal reflux' is denied.

References

1   Bähren W, Lenz M, Porst H, Wierschin W: Side-effects, complications and contra-indications of percutaneous sclerotherapy of the internal spermatic vein for the treatment of idiopathic varicoceles (in German). Fortschr Röntgenstr 1983;138: 172–179.
2   Breun H, Jecht E, Schrott KM: Varikozele und distaler Reflux. Urologe A 1982;21: 345–347.
3   Comhaire F, Kunnen M: Selective retrograde venography of the internal spermatic vein: A conclusive approach to the diagnosis of varicocele. Andrologia 1976;8:11–24.

4   Gall H, Bähren W: New aspects in the pathogenesis and treatment of spermatogenetic failure in varicocele; in Orfanos CE, Stadler R, Gollnick H (eds): Dermatology in Five Continents. Proceedings of the XVII. World Congress of Dermatology, May 24–29, 1987. Berlin, Springer, 1988, pp 438–445.
5   Greenberg SH, Lipshultz LJ, Morganroth J, Wein AJ: The use of the Doppler stethoscope in the evaluation of varicoceles. J Urol 1977:117:296–298.
6   Sigmund G, Bähren W, Gall H, Lenz M, Thon W: Idiopathic varicoceles: Feasibility of percutaneous sclerotherapy. Radiology 1987;164:161–168.
7   Zeitler E, Jecht E, Richter E, Seyferth W: The treatment of male infertility by selective catheterization of the spermatic vein. Fortschr Röntgenstr 1980;132:294–300.

Dr. H. Gall, Department of Dermatology, Bundeswehrkrankenhaus, Oberer Eselsberg 40, D-W–7900 Ulm (FRG)

Colpi GM, Pozza D (eds): Diagnosing Male Infertility.
Prog Reprod Biol Med. Basel, Karger, 1992, vol 15, pp 148–152

Plasma Immunoreactive Beta-Endorphin, Cortisol and Dehydroepiandrosterone Sulfate in Infertile Males with Left Varicocele

M. Mariani[a], *M. Polvani*[b], *A. Casarico*[c], *L. Calcagno*[c], *L. Gavazzi*[c]

[a] Department of Public Health and Biostatistics, University of Pisa;
[b] Radioimmunoassay Laboratory USL 3, Versilia Lucca;
[c] Division of Urology, Galliera Hospitals, Genova, Italy

Several theories have been proposed to explain the specific pathophysiology of testicular dysfunction in men with variococele. Reflux of hormonal substances from the suprarenal gland into the renal vein and left spermatic vein with possible toxic effects on the gonads and spermatozoa is one theory [1–5]. To verify this hypothesis, we have assayed plasma immunoreactive beta-endorphin, cortisol (C) and dehydroepiandrosterone sulfate (DHEA-SO$_4$) in plasma samples of infertile patients with left varicocele, collected during percutaneous scleroembolization at the levels of the caval, renal and spermatic veins [6].

Materials and Methods

Fourteen patients with left varicocele underwent computerized Doppler flowmetry as well as transfemoral retrograde phlebography under local anaesthesia (Mepivacaine HCl 2%, 5–10 ml) according to the method of Seldinger [7]. The femoral vein is entered below the inguinal ligament, and a 9-Fr arterial sheath is inserted and taped securely (the sheath is flushed continuously with saline through a side port to maintain patency). A wire-guide is crossed along the caval vein enabling the insertion of a 7.3-Fr curved catheter. Using phlebography, we examine the left external iliac vein for any signs of iliac-cremasteric reflux, the cause of type II and III varicocele (following Coolsaet), and the left renal vein to bring out the reno-spermatic reflux, the cause of type I varicocele.

A first venous blood plasma sample is collected from the inferior vena cava below the renal vein. Once the renal vein is catheterized, a second blood plasma sample is collected. After a superselective internal spermatic vein catheterism, a third venous blood

Table 1. Cortisol values

Subject No.	Varicocele type	Plasma concentrations of cortisol, μg%		
		cava vein	renal vein	spermatic vein
1	I	18.319	15.489	22.728
2	I	20.510	13.159	12.879
3	I	10.554	9.054	10.697
4	I	1.609	0.351	1.547
5	I	2.132	1.318	0.759
6	I	13.586	7.835	5.771
7	I	126.069	15.052	16.542
8	II–III	8.760	6.712	
9	II–III	16.446	15.221	
10	II–III	17.990	17.385	
11	II–III	16.841	13.611	
12	II–III	27.513	23.917	
13	II–III	26.284	15.221	
14	II–III	13.856	9.849	
Mean		23.0	11.7	11.0
SD		30.0	6.0	8.0
SE		8.00	2.0	3.0
Variation coefficient		133.0	53.0	70.0

sample is collected. Once the reno-spermatic reflux has been brought out, and the morphology and the calibre of the internal spermatic vein is examined, the possibility of embolising it with a detachable silicone balloon (as proposed by Halden and White) and/or sclerotizing it, is then evaluated. For beta-endorphin determination, we use the 'Allegro B-endorfin Kit' of Nichols Institute Diagnostics, San Juan Capistrano, Calif., USA.

The determination of beta-endorphin is performed on EDTA + aprotinin plasma in order to minimize peptide degradation. In every sample we also assayed cortisol and DHEA-SO$_4$ by RIA.

Results

Positive phlebography revealed 7 patients with type I varicocele while negative phlebography and iliac-cremasteric reflux revealed 7 patients with types II and III varicocele. All cortisol (table 1) and DHEA-SO$_4$ (table 2) values were in the normal range for every examined region, except

Table 2. DHEA-SO$_4$ values

Subject No.	Varicocele type	Plasma concentrations of DHEA-SO$_4$, µg%		
		cava vein	renal vein	spermatic vein
1	I	148.334	117.679	115.002
2	I	243.842	170.645	175.923
3	I	166.762	169.092	143.884
4	I	111.439	57.540	94.409
5	I	87.294	95.243	36.466
6	I	94.889	96.934	63.560
7	I	248.578	213.119	169.838
8	II–III	164.631	234.532	
9	II–III	218.330	210.701	
10	II–III	126.905	140.705	
11	II–III	140.040	127.767	
12	II–III	487.846	370.828	
13	II–III	321.938	210.842	
14	II–III	196.444	139.614	
Mean		196.9	165.1	126.2
SD		106.0	80.0	60.0
SE		28.0	22.0	21.0
Variation coefficient		54.0	49.0	47.0

for 1 patient with a high cortisol value (≥ 100 µg%). On the other hand, beta-endorphin-like immunoreactivity has very high values in more than 50% of the cases, and a concentration gradient can be seen from the renal to the spermatic vein (table 3).

Discussion

The concentrations are not distributed in a normal gaussian curve, but it is possible to postulate the presence of two subject populations: one of patients with low and one of patients with high plasma concentrations of beta-endorphin. This fact may be in relation to the stress response, so it is possible to rename the subjects with low beta-endorphin values as 'stress nonresponders' and those with high values as 'stress responsers'. The stress of catheterism could stimulate the secretion of beta-endorphin as well as

Table 3. Beta-Endorphin values

Subject No.	Varicocele type	Plasma concentrations of beta-endorphin, pg/ml		
		cava vein	renal vein	spermatic vein
1	I	89.967	78.294	36.897
2	I	38.161	34.950	35.819
3	I	109.332	118.198	100.055
4	I	36.010	36.205	35.864
5	I	35.748	34.748	34.366
6	I	39.177	38.987	37.755
7	I	159.401	89.516	74.781
8	II–III	39.817	39.635	
9	II–III	99.616	99.741	
10	II–III	356.914	374.671	
11	II–III	39.597	36.589	
12	II–III	159.941	122.897	
13	II–III	48.548		
14	II–III	71.832	45.020	
Mean		94.6	88.4	50.8
SD		87.0	92.0	26.0
SE		23.0	25.0	10.0
Variation coefficient		92.0	104.0	51.0

ACTH which stimulate an increase in glycocorticoid and catecholamine suprarenal gland secretion. These substances have been known to be harmful for the gonads [8–10]. In the future, we want to see if these two types of patients are also different in relation to pregnancy rate after surgical correction of varicocele. This study with a few cases and a high standard deviation due to the presence of a patient with very high beta-endorphin-like immunoassay helps to validate the hypothesis, supported by previous experiences with opioid antagonists, that opioid endogenous compounds are involved in infertility.

References

1   Cohen J: Sterilità e subfertilità maschile. Rome, Masson Italia, 1980.
2   MacLeod J: Seminal cytology in the presence of varicocele. Fertil Steril 1965;16: 735.

3 Mariani M: Dati preliminari sull'importanza del trattamento con naloxone per l'indicazione alla correzione chirurgica del varicocele sinistro in soggetti affetti da infertilità. Atti V Congr Naz Soc Ital Androlog. Roma, Acta Medica, 1987, pp 287–300.

4 Mariani M: The use of an opiate receptor blocking agent. A new therapy in human asthenozoospermia. J Androl 1985;6:138.

5 Mariani M: Ruolo dei recettori per gli oppioidi endogeni del tratto uro-genitale maschile nella terapia delle flogosi prostato epididimarie associate ad astenozoospermia; in Giorgino R, Abbaticchio G (eds): Attualità in Andrologia. Bologna, Monduzzi, 1985, pp 113–118.

6 Belgrano E, Trombetta C, Spano G, Tedde A, Deiana I: Terapia percutanea del varicocele. Atti V Congr Naz Soc Ital Androlog. Rome, Acta Medica, 1987, pp 287–300.

7 Casarico A, Pasquini P, Calcagno L, Gavazzi L, Damiani S, Boesmi R, Parodi M, Romagnoli G, Barbieri V: Embolizzazione della vena spermatica interna con palloncino staccabile: una nuova possibilità nella terapia percutanea del varicocele. Acta Urol Ital 1989;1:45–50.

8 Mariani M: Naloxone. Roma, Biblioteca Storia Patria, 1985.

9 Talati JJ, Islahuddin M: The clinical varicocele in infertility. Br J Urol 1988;61:354–358.

10 Caldamone AA, AjJuburi A, Cockett ATK: The varicocele: Elevated serotonin and infertility. J Urol 1980;123:683–685.

Dr. M. Mariani, Dipartimento di Sanità Pubblica e Biostatistica,
Università di Pisa, I–56100 Pisa (Italy)

Colpi GM, Pozza D (eds): Diagnosing Male Infertility.
Prog Reprod Biol Med. Basel, Karger, 1992, vol 15, pp 153–157

Genitourinary Tuberculosis in Males and Its Relation to Infertility

S. Tellaloğlu, H. Ander, A. Kadloğlu, A. Öner, A.Y. Müslümanoğlu[1]
Department of Urology, Medical Faculty of Istanbul, Turkey

Pulmonary and nonpulmonary tuberculosis still remain serious problems, especially in third world countries. The World Health Organization has estimated that throughout the world there are ten million new cases of all forms of tuberculosis [1].

The overall tuberculosis rate in Turkey is higher than those in developed countries (3.5‰/0.13‰) [1, 2]. *Mycobacterium tuberculosis* reaches the genitourinary organs by the hematogenous route from the lungs. The primary site may not be symptomatic or apparent [3].

Kidney and ureter, bladder, prostate and seminal vesicles, spermatic cord, epididymis and testis become involved by different routes of dissemination [1, 3].

The first clue for genital tuberculosis is usually epididymal findings. The disease starts in the globus minor of this organ because of its greater blood supply [4]. Urinary tuberculosis can be associated with tuberculosis of the epididymis. Lattimer and Wechsler [5] reported that in 40% of 150 cases of urinary tuberculosis occurrence of epididymitis has been of diagnostic value.

The ducts of the involved epididymis become occluded and if this is the case, bilateral azoospermia results [1]. The infertility rate of genital tuberculosis has been found to be 2.2% in our department, where 352 infertile men were investigated.

---

[1] We thank Dr. Gürhan Özdemir (Medical Faculty of Ankara) for his collaboration in this study.

Table 1. Age distribution

Age, years	Number of patients
< 22	2
22–25	8
26–30	11
31–35	9
36–40	4
41–45	4
46–50	4
51–55	2
56–60	4
61–65	2
> 65	1

Material and Methods

Between 1980 and May 1990, 51 cases of genitourinary tuberculosis were examined in our clinic. The ages of these patients are shown in table 1.

The patients were evaluated with history, physical examination, roentgenological (chest, IVU) and ultrasonographical examination (7.5 mHz transducer), hemogram, ESR, urine and ejaculate culture (5 ×), secretion (of the fistula material) culture, and PPD (skin test). Endoscopy and biopsy were performed in some cases and sperm counts were obtained in 29 of 51 patients. If there is abnormality in spermiograms they were repeated every 3 months.

Results

The initial complaints of the 51 patients were as follows: scrotal swelling, testicular pain, scrotal fistula (positive culture in 1 case), micturition disorders, hematuria and hemospermia.

Seven patients complained of infertility and 22 patients had previous history of tuberculosis with various localization.

Among the 51 patients 23 had renal and the remaining 28 had genital tuberculosis. Three patients in the renal tuberculosis group had associated epididymal findings. In the patients of the genital tuberculosis group, the epididymis was always affected. The epididymal findings were left sided in 18, right sided in 5 and bilateral in 5 cases. Rectal palpation in 6 patients suggested tuberculosis of the prostate and in 1 patient the prostate and

Table 2. Improvement in semen analysis of 4 patients

Case	Sperm count and motility	
	initial	during treatment
1	$8\times10^6/40\%$	$23\times10^6/40\%$
2	$8\times10^6/80\%$	$35\times10^6/90\%$
3	$10\times10^6/40\%$	$86\times10^6/73\%$
4	98×10^6/motility cannot be evaluated significant hemospermia	$110\times10^6/86\%$ hemospermia disappeared

seminal vesicles were affected. Seven patients underwent epididymectomy. Six of these operations were performed between 1980 and 1986 and the other one in 1990.

Sperm counts were performed in 29 patients; 13 patients showed azoospermia. The other sperm counts ranged from 8×10^6 to normozoospermia. In one patient with severe hemospermia, motility of the sperm could not be evaluated.

Four patients showed improvement in semen analysis after medical treatment (table 2). This improvement was in sperm concentration in 2 patients, and in sperm concentration and motility in 1 case. Severe hemospermia disappeared in 1 patient after medical treatment for 1 year.

Discussion

In spite of a general decline in cases of tuberculosis, the urologist must be aware of genitourinary tuberculosis, because this can develop in 2–20 (mean: 8) years after the initial disease.

The symptoms of genitourinary tuberculosis varied in our series. Genital tuberculosis was always associated with epididymal findings according to the data in the literature [1, 5].

Tuberculosis is usually a disease of the young. It is estimated that 71% of the infected population is below the age of 20. Twenty-eight patients are in the 22–35 age group in our series (54.9%).

Scrotal ultrasonography is a noninvasive method in the diagnosis of epididymitis and its complications such as abscess or fistula formation [6].

A 5.0- or 7.5-mHz transducer is enough for evaluating epididymal echogeneity and scrotal content [6].

Reports about [7] the association between epididymitis and the development of antisperm antibodies point out that these antibodies can persist for 3 years if the antigenic impulse remains. Sperm count and motility can be affected by antisperm antibodies. The alterations in spermiogram due to antibodies should be regarded during fertility investigation.

Obstructed epididymis ducts also result in changes in the spermiogram. Jimenez-Cruz et al. [8] reported 100% cytomorphological and biochemical alterations in the spermiograms of the patients in the genital tuberculosis group, while this rate is 75% in the urinary tuberculosis group [8].

Azoospermia can develop due to obstructed epididymis, vas deferens and ejaculatory ducts. In some of these cases, sperm can reappear which is explained by some authors as recanalization [5]. Therefore, medical therapy should be the treatment of choice.

Thirteen patients showed azoospermia and bilateral epididymal findings were noticed in 5 patients. Clinical and ultrasonographical epididymal findings did not correlate with semen analysis in our series. Contralateral epididymis can be attacked without clinical and ultrasonographical clues. Antisperm antibody formation could also be responsible for the changes in the spermiogram.

Another aspect of tuberculosis causing infertility is systemic secondary amyloidosis, which occurs after chronic infection of tuberculosis. In a study at the Medical Faculty of Ankara testicular biopsy was taken in 30 cases for the diagnosis of amyloidosis. Six of these were of tuberculous origin and four had azoospermia [Gürhan, personal commun.].

In conclusion, tuberculosis is still a serious health problem in some countries. Genitourinary tuberculosis can develop after the primary disease and epididymis is the target organ in genital tuberculosis. Obstructive azoospermia and therefore infertility can develop in some cases, but medical therapy should be the choice of the treatment since recanalization can occur.

References

1　Cow JG: Genitourinary tuberculosis; in Walsh DC, Gittes RE, Perlmutter AD, Stamey TA (eds): Campbell's Urology. Philadelphia, Saunders, 1986, pp 1037–1069.
2　Akkaynak S: Epidemiology of the Tuberculosis; in Akkaynak S (ed): Tuberculosis. Ankara, Ayyildiz Company, 1986, pp 54–64.

3   Tanagho EA: Specific infections of the genitourinary tract; in Tanagho EA, McAn-
      inch JW (eds): Smith's General Urology. Norwalk, Appleton & Lange, 1988, pp
      246–261.
4   Macmillan EW: Blood supply of the epididymis in man. Br J Urol 1954;26:60.
5   Lattimer JK, Wechsler M: Genitourinary Tuberculosis; in Harrison JH, Gittes RF,
      Perlmutter AD, Stamey TA (eds): Campbell's Urology. Philadelphia, Saunders,
      1978, p 563.
6   See WA, Mack LA, Krieger JN: Scrotal ultrasonography: A predictor of complicated
      epididymitis requiring orchiectomy. J Urol 1988;139:55–56.
7   Ingerslev HJ, et al: A prospective study of antisperm antibody development in acute
      epididymitis. J Urol 1986;136:162–164.
8   Jimenez-Cruz JF, De Cabezon JS, Soler-Rosello A, Sole-Bolcells F: The spermio-
      gram in urogenital tuberculosis. Andrologia 1979;11:67–70.
9   Özdemir Gürhan: Personal communication.

Dr. S. Tellaloğlu, Department of Urology, Medical Faculty of Istanbul,
34390 Istanbul (Turkey)

Colpi GM, Pozza D (eds): Diagnosing Male Infertility.
Prog Reprod Biol Med. Basel, Karger, 1992, vol 15, pp 158–163

Electroejaculation: Diagnosis and Treatment of Ejaculatory Dysfunction

*Y. Shotland, M. Direnfeld, S. Goldman, J. Kalderon, H. Abramovici,
A. Lurie*

Departments of Urology and Gynecology, Carmel Hospital, Haifa, Israel

Ejaculation is a complex process involving emission, antegrade ejaculation and bladder neck closure [1]. The efferent stimulations are mediated through a center located in the sympathetic ganglion chains at level T_{11}-L_3 [2]. Anejaculation is a disorder that occurs rarely in the general population. Any disruption of neurologic pathways or functional ejaculatory failure can interfere with the ejaculatory process. Sex therapy, classic psychotherapy, vibratory stimulation, and alpha-adrenergic sympathomimetic agents failed to induce emission and ejaculation of many patients. Most of the men with upper spinal cord injuries remained refractory to those treatments. The method of electrostimulation to intact neurologic neurons that induce emission and ejaculation remains the only hope for these couples [3–5]. Sequential stimulations and use of a special catheter positioned at the bladder neck improve volume, mean concentration and motility of sperm cells. Semen thus obtained can be used immediately for artificial insemination or with combination of the IVF technique.

Material and Methods

From April 1989 to April 1990, 27 anejaculate males underwent 43 procedures to obtain sperm volume by transrectal electroejaculation. The indications were: paraplegia in 13, quadriplegia in 5, psychogenic anejaculation in 3, multiple sclerosis in 2, diabetes mellitus in 2, cordoma of the spinal cord in 1, and anejaculation postbilateral hernior-

Table 1. Indication of electroejaculation

Indication	Patients (n = 27)
Paraplegia	13
Quadriplegia	5
Psychogenic anejaculation	3
Multiple sclerosis	2
Diabetes mellitus	2
Cordoma of spinal cord	1
Fibrosis postbilateral herniorrhaphy	1

rhaphy in 1 patient (table 1). The age range was 23–47 years (mean 32) and the time interval from injury was 1–28 years (mean 11). In 18 males (66.6%) the reason for anejaculation was trauma, half of them during military service. The other 9 males complained of anejaculation due to neurogenic and psychogenic impairment.

A urologic, neurologic, psychologic and sexual history of the male was obtained by interview. Medical evaluation included routine blood test, blood pressure, urine culture, intravenous pyelography or ultrasonography and Rigi-scan testing. All patients signed a consent form appropriate to the government's requirement on standards for human investigation.

All patients underwent urodynamic evaluation including cystometrography, electromyography and urethral pressure profile studies to prove low pressure in proximal and distal sphincter area responsible for retrograde ejaculation. To prevent retrograde ejaculation during electrostimulation, we used a urodynamic catheter (UPC-10 Life-Tech) with an inflatable balloon (3 cm^3) positioned at the bladder neck. The outer diameter of the catheter is very thin (3 mm) and causes the sperm fluid to move anteriorly without difficulty (fig. 1). The catheter was used in paraplegic and quadriplegic patients following sphincterectomy as well as in patients complaining of retrograde ejaculation due to diabetes mellitus, multiple sclerosis and others. The electroejaculation equipment contains a stimulator, thermoindicator and several probes that vary in diameter. The rectal probe has three ventral electrodes for stimulation of the maximum mucosal surface and a built-in thermosensor for monitoring temperature to prevent mucosal injury.

The patient is placed in the lithotomy position and catheterization is done to evacuate residual urine before the procedure. A sperm medium was used to flush the bladder before stimulation. The electroejaculation probe is placed in the rectum and gradually electric current stimulations are done with increasing voltage. After antegrade expulsion of the ejaculate and termination of the stimulations, the bladder is catheterized again to check the presence of sperm cells in the urine (table 2).

We began initially with stimulations of 1 V and proceeded up to a maximum of 25 V, with 1 V stimulus increments. All patients were monitored for change in ECG and blood pressure, at 1-min intervals during each stimulation procedure. Before stimulation we use sublingual nifedipine to minimize the elevated blood pressure associated with autonomic hyperreflexia in the upper motor neuron lesion [5].

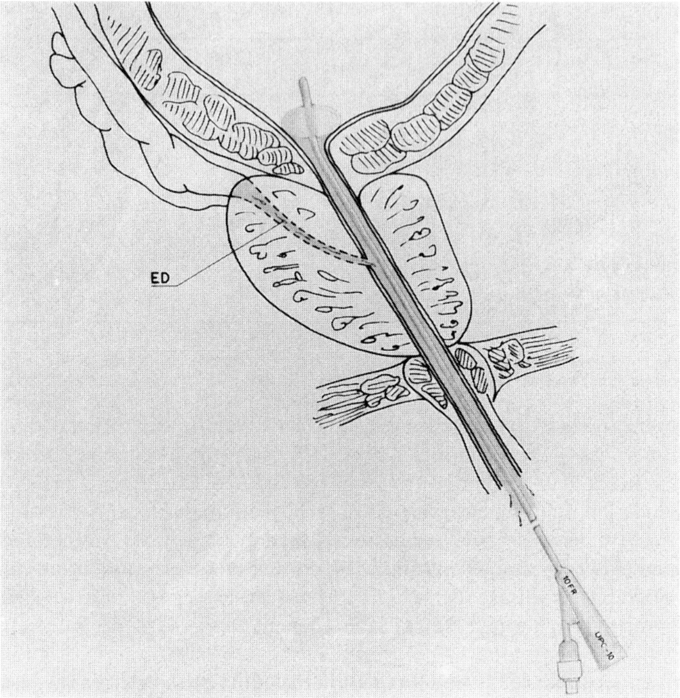

Fig. 1. Insertion of intraurethral catheter. ED = Ejaculatory duct.

Results

Before electroejaculation no patient obtained sperm fluid by masturbation or by mechanical stimulation. Only 3 males (12%) had spontaneous ejaculation during night and in 7 males (26%) sperm cells were found in the urine at certain points in the past. After stimulation we successfully obtained sperm fluid in the first session in 22 males (81%) with improvement of volume, concentration and motility in further stimulations. We failed to get sperm volume in only 5 patients; in 3 males due to severe hypospermatogenesis that was proved by biopsy of testis. In 2 other patients, the diagnosis of obstruction was done by failure to get sperm cells after electrostimulation, confirmed later by normal spermatogenesis after biopsy and normal hormonal levels. In 31% of patients we got sperm volume only anteriorly. Retrograde ejaculation after stimulations was

Table 2. Technique of electrostimulation

Position: lithotomy
Most of patients: no anesthesia
Pre- and postcatheterization
Flushing the bladder with sperm medium
Time interval of stimulation: 3–5 min
Transrectal, gradually, low intensity stimulations
No complications
Use of special catheter, preventing retrograde ejaculation
Artificial insemination at monthly basis during ovulation

Table 3. Results of electroejaculation

	First cycle	Sequential cycles	After swim-up (n = 8)
Volume, cm^3	0.2–7 (3.7)	4.8 (30%)[1]	0.3
Concentration, $\times 10^6$	1–200 (33)	47 (45.4%)	0.5–1
Motility, %	0–40 (13)	15 (25%)	90

[1] Improvement of parameters by percentage in sequential cycles.

achieved in 69% of males alone, or with combination with antegrade ejaculation.

To date, all of the males underwent 43 stimulation cycles: 18 males one cycle, 5 males two cycles, 1 male three cycles and 3 males four cycles of stimulations. Mean volume of sperm fluid in the first cycle was 3.7 cm^3 (range 0.2–7 cm^3) and increased to mean 4.8 cm^3 in further stimulation. In patients with retrograde and antegrade ejaculations the majority of volume was of retrograde origin. The mean concentration of sperm cells per milliliter was $33 \cdot 10^6$ (range $1 \cdot 10^6 – 200 \cdot 10^6$) and increased to mean $47 \cdot 10^6$ in sequential stimulations. The motility of sperm cells was very poor, mean 13% (range 0–40%) and increased only to 16% in further stimulation. After processing of the semen and after the swim-up procedure, the motility increased up to 90% but still without initiating pregnancy (table 3). With the use of a catheter to prevent retrograde ejaculation, the mean

Table 4. Results of electroejaculation with use of a catheter

Etiology	Level of injury	Ejaculate volume, cm^3	Concentration, $\times 10^6$	Motility, %
Psychogenic	cortex	3–4	12–30	5–15
Diabetes mellitus	periph. neurop.	3–3.5[1]	40–60	0–30
Quadriplegia	C_6–C_7	3–5[1]	5–6	0–10
Paraplegia	D_{12}	3–4.5[1]	30–40	0–15
Paraplegia	D_{12}–L_1	2.5–13[1]	105–130	10–20

[1] Use of catheter to prevent retrograde ejaculation.

volume of ejaculate doubled in the second session. The mean sperm count per milliliter improved by 24% and the mean motility of the sperm cells increased from 3.8 to 21% in the second session as compared to the first one (table 4).

Discussion

Electroejaculation is a well-recognized technique which has been used to induce emission in anejaculate males. Unfortunately, the quality and motility of sperm cells obtained by this method is very poor. Systemic and local factors like neurogenic impairment, hyperthermia, chronic infection and nondrainage are the main reasons that influence the fertility of these males. In a very selective group of patients this noninvasive method can be used for diagnosis of obstruction of ejaculatory ducts. In 2 patients in our study we failed to obtain sperm cells after stimulation, most probably due to fibrosis after irradiation and operation. Normal hormonal levels and normal spermatogenesis shown at light microscope confirmed obstruction and prevented unnecessary vasography.

Our recent modification to prevent retrograde ejaculation by use of a thin catheter positioned at the bladder neck improved significantly volume, count and motility of sperm cells. In future the semen ejaculate by this method can be cryopreserved in groups of psychogenic anejaculate males and in early spinal shock patients who desire to be fertile. Further improvement in processing of sperm cells, swim-up technique and advancement in the field of male infertility could hopefully increase fertility of those men who wish to father children.

References

1 Newman HF, Reiss H, Northrup ID: Physical basis of emission, ejaculation and orgasm in males. Urology 1982;19:341–350.
2 Thomas AJ Jr: Ejaculatory dysfunction. Fertil Steril 1983;39:445.
3 Brindley GS: Electroejaculation: Its technique, neurological implications and uses. J Neurol Neurosurg Psychiatry 1981;44:9–18.
4 Bennett CJ, Seager SW, Vasher E, McGuire EJ: Sexual dysfunction and electroejaculation in men with spinal cord injury. Review. J Urol 1988;139:453–457.
5 Murphy JB, Lipshultz LI: Abnormalities of ejaculation. Urol Clin North Am 1987; 14:583–596.

Dr. Y. Shotland, Department of Urology and Gynecology, Carmel Hospital, 34404 Haifa (Israel)

Colpi GM, Pozza D (eds): Diagnosing Male Infertility.
Prog Reprod Biol Med. Basel, Karger, 1992, vol 15, pp 164–177

Autosomal Rearrangements and Reproductive Failure in Man[1]

Y. Rumpler, O. Gabriel-Robez, C. Ratomponirina[2]

Institut d'Embryologie, Université Louis-Pasteur, Faculté de Médecine,
Strasbourg, France

Many studies have shown that sexual chromosomal rearrangements generally give rise to sterility in humans. Autosomal rearrangements are also able to reduce fertility of the carriers, not only because of abortions due to the fertilization of unbalanced gametes, but also because of a strong reduction in gametogenesis. Thus, Robertsonian or reciprocal translocations are found in excess of up to 10-fold in men presenting with hypofertility, compared to the general population [9, 11]. Meiotic studies undertaken have shown anomalies such as a reduction in the number of chiasmata in hypofertile men [8], or an association between the sexual bivalent and the rearranged chromosomes in mice [10] and in man [26].

By using the electron microscope, it was found that the rearranged chromosomes show a delay in pairing [12, 29, 30] and that the unsynapsed elements often come into contact with the sexual bivalent [5, 13, 17, 30]. Taking into account these results and considering that female carriers of chromosomal translocations do not show the same reduction in fertility as do males [13], most authors have proposed that male sterility was linked to an interaction between the rearranged chromosomes and the sexual XY

[1] Supported by a grant from the Foundation pour la Recherche Médicale Française.

[2] We thank Mr. J.L. Maetz for his technical assistance, Mr. G. Cadiou for his collaboration in photographic work, Mrs. M. Lavaux for secretarial assistance, and Mr. J. Anderson for correcting the English version of the manuscript.

bivalent [23]. However, the meiotic study of several sterile human males carrying chromosomal rearrangements, not associated with the sexual bivalent, present data in favor of another hypothesis, namely that the failure in synapsis is by itself deleterious to gametogenesis, without any mediation of the XY chromosomes [1, 4, 6, 31]. In this paper, different hypotheses will be discussed, taking into account our laboratory's results.

Material and Methods

An electron microscopic study of meiosis was performed in 8 cases of chromosomal rearrangements in men presenting with hypofertility or sterility: 1 case of Robertsonian translocation, 4 cases of reciprocal translocations and 3 cases of pericentric inversions (table 1). Testicular material, obtained in the propositus from biopsy under local anesthesia, was partly used for routine pathological examination and for electron microscopic study. Andrological investigations were undertaken in all cases and showed oligospermia or azoospermia. Meiotic preparations were obtained using the microspreading technique of Moses et al. [28]. The microspread spermatocytes were stained with $AgNO_3$ [2] or phosphotungstic acid [7]. The best preparations were photographed at a magnification of 3,000 using a Philips 60 kV microscope.

Results

Synaptonemal Complexes of the Normal Chromosomes
As has been described by Solari [33], synapsis of the metacentric autosomes starts at zygotene stage at the level of the 2 telomeric regions and progresses toward the centromere. For the chromosomes X and Y, pairing starts at the early pachytene stage, on a short terminal segment, while the 2 other extremities tend to come together. The unsynapsed part of chromosomes X and Y becomes hyperchromatic and of irregular shape. At the end of the pachytene stage, the sexual bivalent takes on an inrregular weblike shape. Considering the morphology of the sexual bivalent, it is possible to determine the chronology of the pachytene stage, as has been shown by Solari [33] (fig. 1).

Synaptonemal Complexes of the Rearranged Chromosomes
The rearranged chromosomes show a different appearance according to the type of chromosomal rearrangement, but in all cases they show a delay in pairing compared to the autosomal normal bivalents.

Table 1. Types of chromosomal rearrangements, sperm count and some characteristics of the synaptonemal complexes

Types of rearrangements	Sperm count	Diag-nosis	Sexual bivalent rearranged autosome association, %	Hetero-synapsis %	References
t(13;14)	173×10^6	S	20		Johannisson et al., 1987 [23]
t(13;14)	1.6×10^6	S	61		Luciani et al., 1984 [26]
t(13;14)	6×10^6	S	40		Gabriel-Robez et al., 1988 [17]
t(14;21)	severe oligospermia	S	75		Rosenmann et al., 1985 [30]
t(14;21)	severe oligospermia	S	75		Rosenmann et al., 1985 [30]
t(14;21)(q13;p13)	azoospermia	S	65	0	Johannisson et al., 1987 [23]
t(19;22)(13.1;q11.1)	azoospermia	S	70	0	Gabriel-Robez et al., 1986 [13]
t(17;21)(p13;q11)	1.9×10^6	S	70	0	Gabriel-Robez et al., 1986 [13]
t(4;13)(q11;q123)	2.5×10^6	S	38	20	in preparation
t(9;15)(p22;q15)	32×10^6	I	42	4.5	Johannisson et al., 1987 [23]
t(9;20)(q34;q11)	10×10^6	I	20	17	Chandley et al., 1986 [5]
t(6;4)(q27;q26)	$84 \times 10^6 – 16 \times 10^6$	F	2	50	in preparation
inv 1(p32;q42)	6×10^6	S		17	Gabriel-Robez et al., 1986 [12]
inv 1(p31;q45)	1×10^6	S		16	Chandley et al., 1987 [6]
inv 1(p32;q42)	azoospermia	S		0	Batanian and Hulten, 1987 [1]
inv 2(p13;q35)	unknown	F		18	Hulten et al., 1987 [20]
inv 6(p22;q22.2)	4×10^6	I		35	Gabriel-Robez et al., 1987 [14]
inv 13(p12;q14)	unknown	F		57	Saadallah and Hulten, 1986 [31]
inv 21(p11.2;q21.2)	12×10^6	F		100	Gabriel-Robez et al., 1988 [15]
t(9;12;13)(q22;22;q32)[1]	54×10^6	F	33	22	Johannisson et al., 1988 [24]
t(9;12;13)(q22;22;q32)[1]	329×10^6	F	12.3	12	
t(9;12;13)(q22;22;q32)[1]	$43 \times 10^6; 4 \times 10^6$	I	2.8	11	

S = Sterile; I = infertile; F = fertile. In the column diagnosis, I means that there are no offspring but that repeated miscarriages occurred. In the column heterosynapsis, immediate heterosynapsis as well as the so-called 'synaptonemal adjustment' of Moses are reported.
[1] Three brothers with the same chromosomal arrangement.

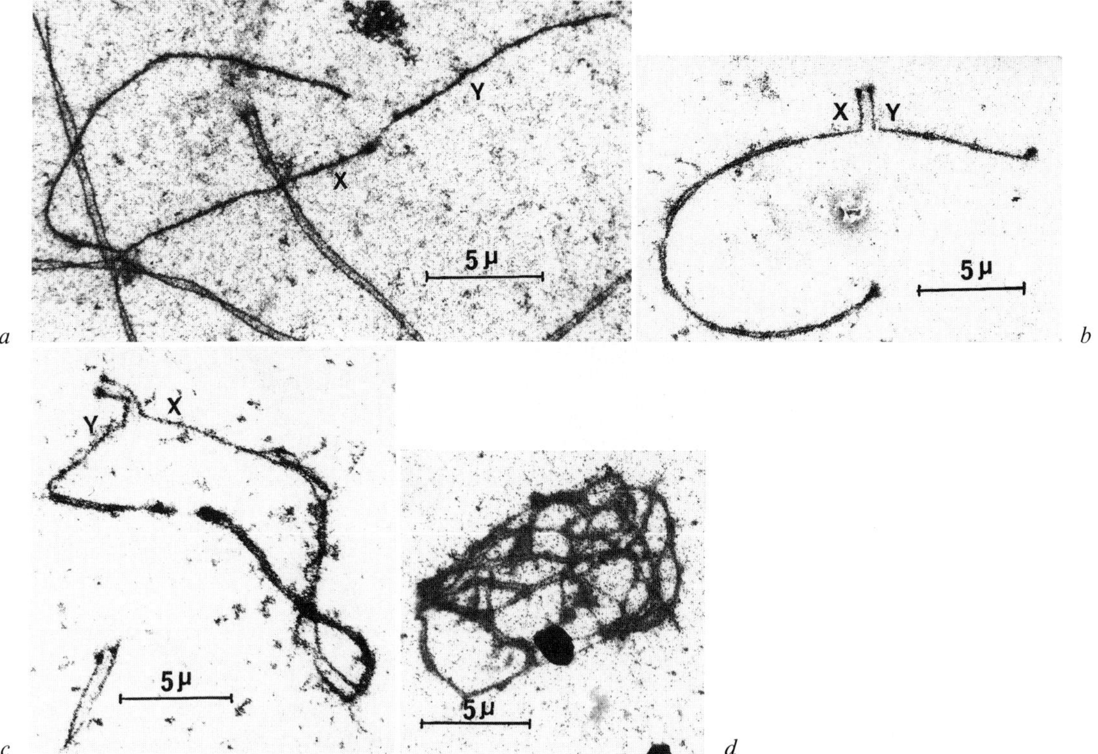

Fig. 1. Partial pachytene nuclei showing progressive pairing and modification of the synaptonemal complexes of the X and Y chromosomes. *a* Late zygotene: the X and Y chromosomes are dark stained and synapsis is not yet initiated. *b* Early pachytene: both sex chromosomes are synapsed on a short part of their length, whereas the two other extremities often come together. *c* Mid pachytene: the unsynapsed parts of the X and Y chromosomes are thickened and of irregular shape. *d* Late pachytene: X and Y take on an irregular web-like shape which is thickened and dark stained.

Robertsonian Translocations. The Robertsonian translocation t(13;14) gives rise to trivalents as the two cases of t(14;21) reported by Rosenmann et al. [30], the two normal acrocentric chromosomes pairing with the translocated metacentric chromosomes (fig. 2a). Synapsis starts at the level of the distal part of the trivalent and progresses toward the kinetochores. The two short arms of the acrocentric chromosomes, which are of heterochromatic composition, remain unsynapsed for a time. In some cells the two

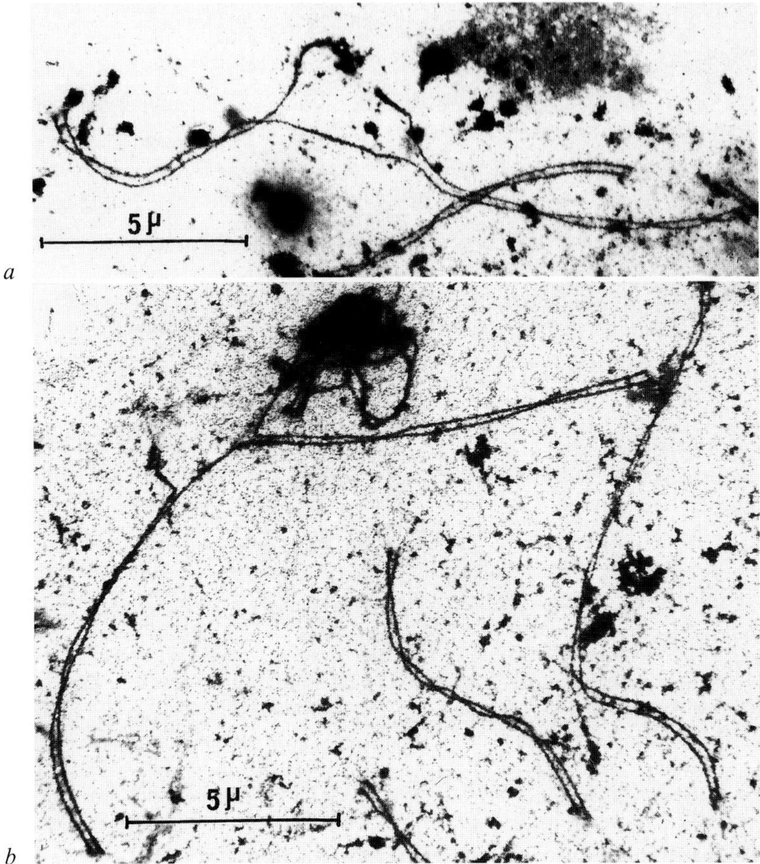

Fig. 2. a Mid pachytene showing the delay in synapsis of the trivalent. *b* Late pachytene showing the free arms often in contact with the sexual bivalent.

short arms will synapse later during the pachytene stage, whereas in others, they will remain unsynapsed. The unsynapsed arms have a hypercondensed aspect and often come into contact with the sexual bivalent (fig. 2b; table 1).

Reciprocal Translocations. There are differences in the pairing of the quadrivalent formed by the reciprocal translocation involving an acrocentric chromosome or not, even if in both cases a delay in pairing occurs.

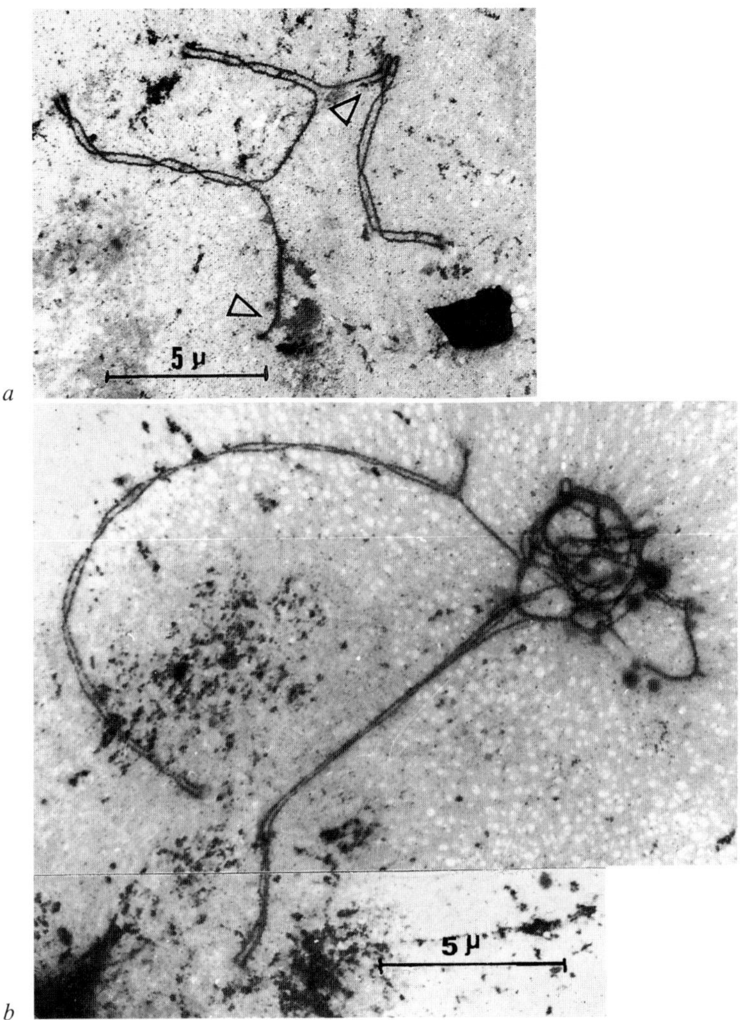

Fig. 3. Microspreading preparations showing: *a* A quadrivalent resulting from the translocation t(4;13). One of the two unpaired proximal parts of chromosomes 13 (arrows) is more stained. *b* The quadrivalent is in close contact with the sex chromosomes. Two arms are visible whereas the two others are so entangled with the XY bivalent that they cannot be distinguished from the XY network.

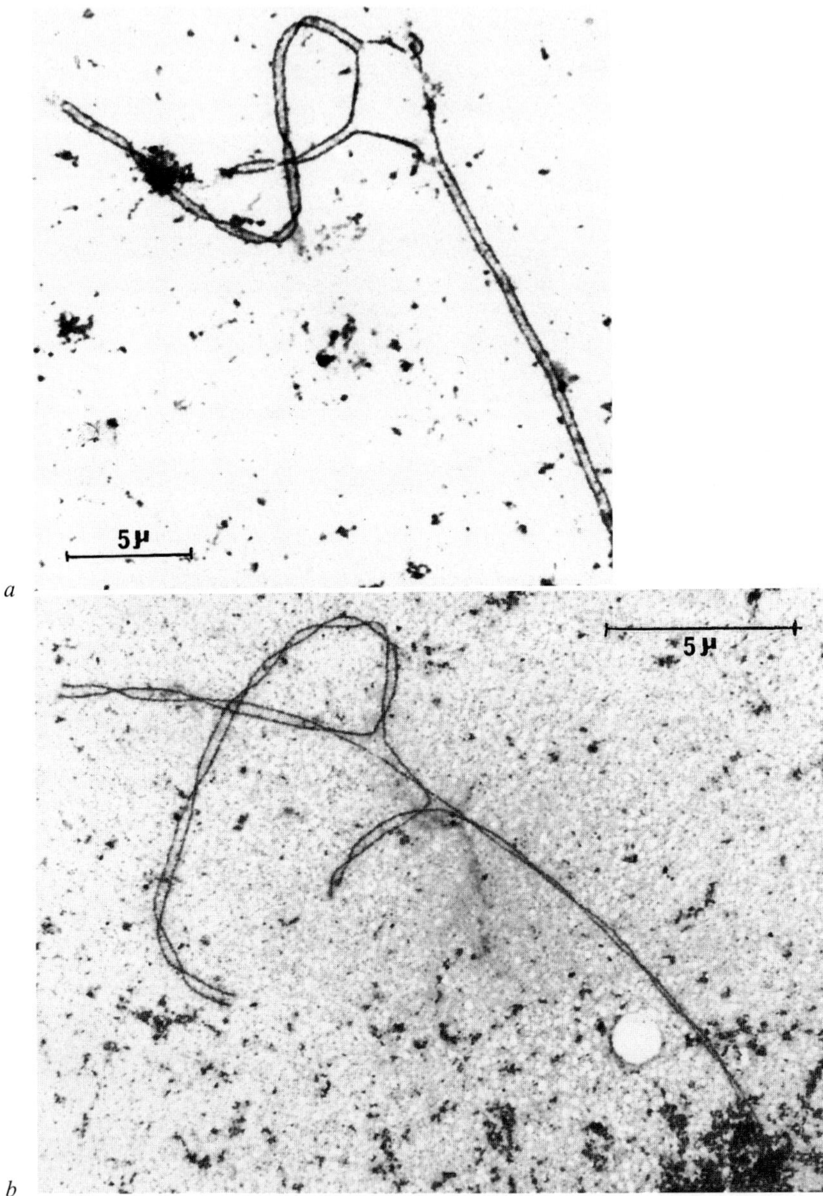

Fig. 4. Microspreading preparations showing the synaptonemal complexes of a translocation t(4;6). *a* The quadrivalent shows a delay in pairing of one of the arms and around the breakpoints. *b* Heterosynapsis of the axes around the breakpoints.

Reciprocal Translocations Involving an Acrocentric Chromosome. Reciprocal translocations t(17;21) t(9;22) reported earlier [13] and the t(4;13) [new case] give rise to a quadrivalent (fig. 3a) but initiation of their synapsis starts most often at the level of 3 of the 4 extremities. The fourth, probably corresponding to the heterochromatic short arms of the acrocentric chromosomes, often presents a delay in pairing, similar to the heterochromatic arms in trivalents. In a large number of cells the unsynapsed arms of the quadrivalent come into contact with the sexual bivalent (table 1) and show morphological modifications. In some spermatocytes, segments of varying lengths of the quadrivalent, probably corresponding to unsynapsed elements are included in the network of the XY bivalent and are indistinguishable from it (fig. 3b). The elements remaining outside the XY network and therefore normally synapsing are of normal staining intensity.

Reciprocal Tanslocations without Acrocentric Chromosome (fig. 4). Reciprocal translocations between nonacrocentric chromosomes t(4;6) [this paper] also give rise to a characteristic quadrivalent as the t(9;20) described by Chandley et al. [5]. Synapsis of the arms of the quadrivalent starts at the four distal extremities in almost all the spermatocytes studied and progresses toward the breakpoints which often show a delay in pairing as in the quadrivalent of the case of t(9;20) reported by Chandley et al. [5]. But in a large number of cells (table 1) a heterosynapsis occurs at early pachytene yet. There is a difference between this case and that reported by Chandley et al. [5], the latter showing a failure in pairing of one of the four arms of the quadrivalent, which often comes into contact with the XY bivalent, whereas in the case reported here, the association is very rare (1/36) (table 1). In both cases, the unpaired arms occasionally show thickening, whether or not they are in contact with the XY bivalent.

Pericentric Inversion. In 2 cases, inversion 1 and inversion 6 reported by Gabriel-Robez et al. [2, 3], synapsis occurs in the two classical steps proposed by Moses et al. [28]. During the first step, homosynapsis occurs. It starts in most cases at the two extremities and progresses along the chromosomes to the inverted segment remaining temporarily unsynapsed. Pairing of this region starts in the middle of the inverted segment and progresses in both directions giving rise to a characteristic inversion buckle in a large number of spermatocytes [14], as has been described in mice by Moses et al. [28] and in man Gabriel-Robez et al. [12] (fig. 5a). The second

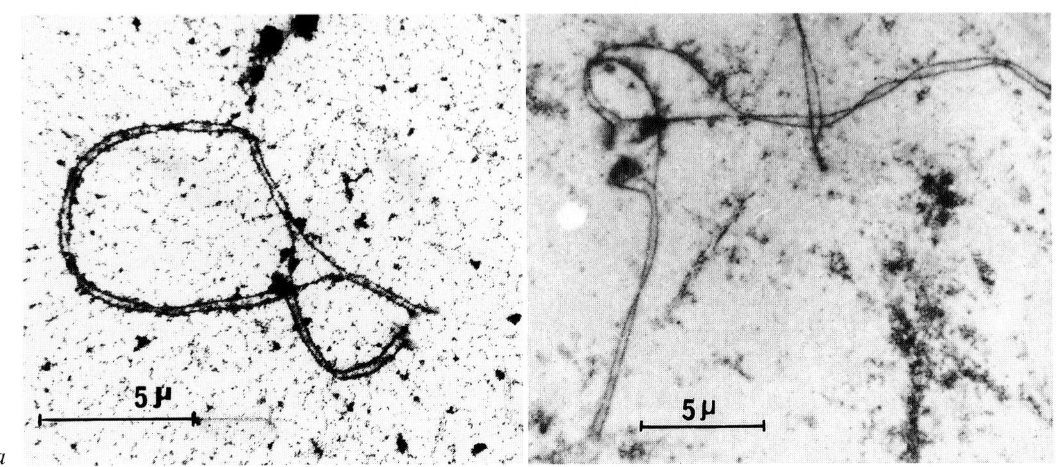

Fig. 5. a Synaptonemal complexes of a pericentric inversion of chromosome 6, showing a characteristic full inversion buckle. *b* Synaptonemal complexes of a pericentric inversion of chromosome 1 showing a thickening of the asynapsed segments corresponding to the buckle.

step, called synaptic adjustment by Moses et al. [28] follows homosynapsis and progressively eliminates the loop leading to a straight heterosynaptic pairing of the inverted segment.

. Even if homosynapsis mainly occurs during early pachytene and heterosynaptic adjustment during mid and late pachytene, a complete synaptic adjustment has been found at early pachytene in some spermatocytes (4/21) of the inversion 6 carrier so that it seems not to have been preceded by an inversion buckle, as reported by Chandley et al. [6] in case of inversion 1. In both cases the asynapsed regions during early pachytene stage often show a hypercondensation with thickening and excrescences either on their entire length or only a part when synapsis has started (fig. 5b).

In the third case concerning an inversion of chromosome 21 (fig. 6) [15], no spermatocyte shows an inversion buckle in the 50 cells analyzed. It appears that immediate heterosynapsis occurred at early pachytene giving rise to a linear pairing. The inversion bivalent can be unambiguously identified by its non aligned kinetochores (fig. 6a, arrows).

In the 3 cases studied in our laboratory, as well as in the cases reported by other authors [1, 6, 20, 31], no association between the inverted chromosome and the sexual bivalent has been observed (table 1).

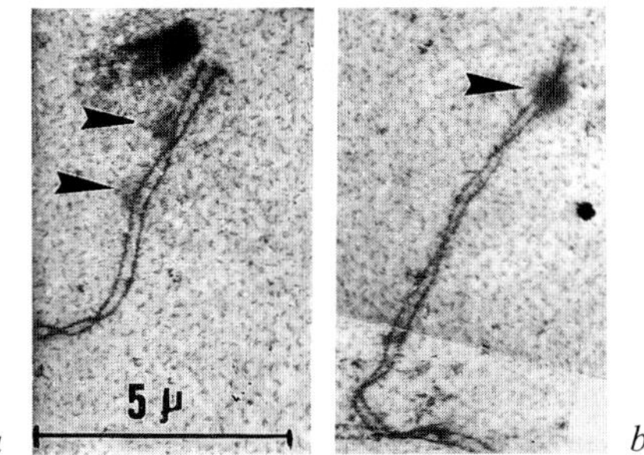

Fig. 6. Synaptonemal complexes of a pericentric inversion of chromosome 21 in mid pachytene. *a* Heterosynapsis of bivalent 21 linearly paired, showing the no aligned kinetochores (arrows). *b* Bivalent 22 from the same cell showing aligned kinetochores.

Discussion

On the basis of the work of Monesi [27] who demonstrated that in spermatocytes the X chromosome inactivation occurs earlier than that of the autosomes during the pachytene stage, as well as the study by Hotta and Chandley [19] who showed that the enzymatic activity related to X-linked genes was more important in carriers of X-autosome translocation than in normal mice, most authors have suggested that the degeneration of spermatocytes I is the result of XY reactivation each time the sexual bivalent comes into contact with the rearranged chromosome [25, 26, 30]. Some authors have assumed that the inactivation of the sexual bivalent could also spread from the gonosomes to the associated autosomes [29, 34]. In favor of this latter hypothesis, Jaafar et al. [21] showed, in an autoradiographic study on spermatocytes of X-16 translocated mice, that there is no global reactivation of the XY bivalent but that the translocated autosomal segment in direct contact with the X chromosome has a reduced transcriptional activity. Thus, it seems that the deleterious interaction between the X chromosome and the autosomes leading to germ cell death is due to a reactivation of some inactivated X-linked genes as well as to a partial transcriptional inactivation of autosomal segments [21]. In accordance with this view, the data reported in table 1 show that the severity of the oligospermia is roughly related to the percentage of spermatocytes exhibiting an association between the sexual bivalent and the rearranged chromosomes.

According to the hypothesis of Burgoyne and Baker [4], the failure in pairing can by itself be a major reason for the destruction of the meiocytes in carriers of chromsomal rearrangements. Indeed, in pericentric inversions as well as in some reciprocal translocations, the rearranged chromosomes not being associated with the XY bivalent, spermatogenic failure could not be attributed to a X-autosome interaction. In most cases studied a delay in pairing of the rearranged chromosomes was found, and in many cells showing this failure of pairing the unsynapsed chromosomal segments presented morphological changes such as thickening and/or appearance of excrescences, leading some authors to propose that cells showing these morphological modifications would be eliminated from the germ line [5, 12]. According to this hypothesis, a normal sperm count has been found in boars carrying different reciprocal translocations with an immediate heterosynapsis in about 95% of the spermatocytes [16, 22] as well as in a limited number of species showing pericentric inversions on the heterozygous state in natural populations [18]. Moreover, in view of the data reported in table 1, there appears to be no direct link between the severity of oligospermia and the proportion of spermatocytes showing a heterosynapsis, despite a lesser degenerescence of the germ cells in the inversion 21 and the tranlsocation t(4;6) carriers in which a heterosynapsis occurred immediately. More data have to be collected before conclusions can be drawn. The recent views of Setterfield [32] seem to privilege the role of asynapsis in sterility.

A final point of interest arises from the fact that Johannisson et al. [24] reported that the same complex chromosomal rearrangement involving reciprocal translocations between 3 autosomes showed the same meiotic anomalies in three related heterozygous males, but one male was fertile, the second subfertile and the third sterile. The differences in fertility could not be explained in view of the similar findings for meiotic pairing.

Conclusion

The use of microspreading techniques for investigations in sterile carriers of chromosomal rearrangements showed chromosomal pairing anomalies during the pachytene stage. Despite the data gathered by different authors, it remains difficult to propose an unequivocal explanatory theory to relate the severity of the gametic impairment to the type and the proportion of meiotic abnormalities observed. Further investigations on larger samples need to be performed, especially in collaborative studies between cytogeneticists and urologists.

References

1   Batanian J, Hulten MA: Electron microscopic investigations of synaptonemal complexes in an infertile human male carrier of a pericentric inversion inv(1)(p32;q42). Regular loop formation but defective synapsis including a possible interchromosomal effect. Hum Genet 1987;76:81–86.

2   Bloom S, Goodpasture C: An improved technique for selective silver staining of nucleolar organizer regions in human chromosomes. Hum Genet 1976;34:199–206.

3   de Boer P, Branje HEB: Association of the extra chromosome of tertiary trisomic male mice with the sex chromosomes during first meiotic prophase, and its significance for impairment of spermatogenesis. Chromosoma 1979;73:369–379.

4   Burgoyne PS, Baker T: Meiotic pairing and gametogenic failure; in Evans CW, Dickison HG (eds): Controlling Events in Meiosis. Proc 38th SEB Symp Reading 1983. Cambridge, Company of Biologists, 1984, pp 349–362.

5   Chandley AC, Speed RM, McBeath S, Hargreave TB: A human 9;20 reciprocal translocation associated with male infertility analyzed at prophase and metaphase I of meiosis. Cytogenet Cell Genet 1986;41:145–153.

6   Chandley AC, McBeath S, Speed RM, Yorston L, Hargreave TB: Pericentric inversion in human chromosome 1 and the risk for male sterility. J Med Genet 1987;24: 325–334.

7   Counce SJ, Meyer GF: Differentiation of the synaptonemal complex and the kinetochore in *Locusta* spermatocytes studied by whole mount electron microscopy. Chromosoma 1973;44:231–253.

8   Dutrillaux B, Guéguen J: Anomalies méiotiques et gamétiques multiples dans un cas de stérilité masculine. Ann Génét 1971;14:49–52.

9   Dutrillaux B, Rotman J, Guéguen J: Chromosomal factors in the infertile male; in de Vere White (ed): International Aspects of Male Infertility. Baltimore, Williams & Wilkins, 1982, vol 4, pp 89–109.

10 Forejt J: X-inactivation and its role in male sterility; in Bennett M, Gropp A, Wolff U (eds): Chromosomes Today. London, Academic Press, 1984, vol 8, pp 117–127.

11 Fraccaro M: Chromosome abnormalities and gamete production in man. Differentiation 1983;23(suppl):S40–S43.

12 Gabriel-Robez O, Ratomponirina C, Rumpler Y, le Marec B, Luciani JM, Guichaoua MR: Synapsis and synaptic adjustment in an infertile human male heterozygous for a pericentric inversion in chromosome 1. Hum Genet 1986;72:148–152.

13 Gabriel-Robez O, Ratomponirina C, Dutrillaux B, Carré-Pigeon F, Rumpler Y: Meiotic association between the XY chromosomes and the autosomal quadrivalent of a reciprocal translocation in two infertile men 46,XY t(19;22) and 46,XY t(17;21). Cytogenet Cell Genet 1986;43:154–160.

14 Gabriel-Robez O, Ratomponirina C, Croquette M, Maetz JL, Couturier J, Rumpler Y: Reproductive failure and pericentric inversion in man. Andrologia 1987;19:662–669.

15 Gabriel-Robez O, Ratomponirina C, Croquette M, Couturier J, Rumpler Y: Synaptonemal complexes in a subfertile man with a pericentric inversion in chromosome 21. Heterosynapsis without previous homosynapsis. Cytogenet Cell Genet 1988;48:84–87.

16 Gabriel-Robez O, Jaafar H, Ratomponirina C, Boscher J, Bonneau J, Popescu CP, Rumpler Y: Heterosynapsis in a heterozygous fertile boar carrier of a 3;7 translocation. Chromosoma 1988;97:26–32.

17 Gabriel-Robez O, Ratomponirina C, Cranz C, Weill A, Bollack C, Rumpler Y: Robertsonian heterozygosity and male sterility. Andrologia 1988;20:463–466.

18 Hale DW: Heterosynapsis and suppression of chiasmata within heterozygous pericentric inversions of the Sitka deer mouse. Chromosoma 1986;94:425–432.

19 Hotta Y, Chandley A: 1982. Activities of X-linked enzymes in spermatocytes of mice rendered sterile by chromosomal aberrations. Gam Res 1982;6:65–72.

20 Hulten M, Saadallah N, Batanian J: Meiotic chromosome pairing in the human male: Experience from surface spread synaptonemal complexes. Chromosome Today 1987;9:218–229.

21 Jaafar H, Gabriel-Robez O, Rumpler Y: Pattern of ribonucleic acid synthesis in vitro in primary spermatocytes from mouse testis carrying a X-autosome translocation. Chromosoma 1989;98:330–334.

22 Jaafar H, Gabriel-Robez O, Ratomponirina C, Boscher J, Bonneau M, Popescu CP, Rumpler Y: Analysis of synaptonemal complexes in two fertile heterozygous boars, both carriers of a reciprocal translocation involving an acrocentric chromosome. Cytogenet Cell Genet 1989;50:220-225.

23 Johannisson R, Löhrs U, Wolff HH, Schwinger E: Two different XY quadrivalent associations and impairment of fertility in men. Cytogenet Cell Genet 1987;45:222–230.

24 Johannisson R, Löhrs U, Passarge E: Pachytene analysis in male heterozygous for a familial translocation (9;12;13)(q22;q22;q32) ascertained through a child with partial trisomy 9. Cytogenet Cell Genet 1988;47:160–166.

25 Lifschytz E, Lindsley DL: The role of X-chromosome inactivation during spermatogenesis. Proc Natl Acad Sci USA 1972;69:182–186.

26 Luciani JM, Guichaoua MR, Mattei MA, Morazzani MR: Pachytene analysis of a man with a 13q;14q translocation and infertility. Behaviour of the trivalent and non-random association with the sex vesicle. Cytogenet Cell Genet 1984;38:14–22.

27 Monesi V: Synthetic activities during spermatogenesis in the mouse. Exp Cell Res 1965;39:197–224.

28 Moses MJ, Poorman PA, Roderick TH, Davisson MT: Synaptonemal complex analysis of mouse chromosomal rearrangements. IV. Synapsis and synaptic adjustment in two paracentric inversions. Chromosoma 1982;84:457–474.

29 Ratomponirina C, Viegas-Péquignot E, Dutrillaux B, Petter F, Rumpler Y: Synaptonemal complexes in some *Gerbillidae:* Probable role of intercalated heterochromatin in gonosome-autosome translocations. Cytogenet Cell Genet 1986;43:161–167.

30   Rosenmann A, Wahrman J, Richler C, Voss R, Persitz A, Goldman B: Meiotic association between the XY chromosomes and unpaired autosomal elements as a cause of human male sterility. Cytogenet Cell Genet 1985;39:19–29.

31   Saadallah N, Hulten M: EM investigations of surface spread synaptonemal complexes in a human male carrier of a pericentric inversion inv(13)(p12q14): The role of heterosynapsis for spermatocyte survival. Ann Hum Genet 1986;50:369–383.

32   Setterfield LA, Mahadevaiah S, Mittwoch U: Pachytene pairing in relation to sperm and oocyte numbers in a malefertile reciprocal translocation in the mouse. Cytogenet Cell Genet 1988;49:293–299.

33   Solari AJ: Synaptonemal complexes and associated structures in microspread human spermatocytes. Chromosoma 1980;81:315–337.

34   Viguié F, Romani F, Dadoune JR: Male fertility in a case of (Y;6) balanced reciprocal translocation. Mitotic and meiotic study. Hum Genet 1982;62:225–227.

Dr. Y. Rumpler, Institut d'Embryologie, Université Louis-Pasteur,
Faculté de Médecine, 11, rue Humann, F–67085 Strasbourg Cedex (France)

Colpi GM, Pozza D (eds): Diagnosing Male Infertility.
Prog Reprod Biol Med. Basel, Karger, 1992, vol 15, pp 178–185

Antisperm Antibodies and Assisted Reproduction

F. Dondero, A. Lenzi, F. Lombardo, L. Gandini

Laboratory of Seminology and Immunology of Reproduction, V Medical Clinic, University of Rome 'La Sapienza', Rome, Italy

Since the first demonstration that a significant number of infertile couples manifested an immunity to sperm [1], the experiments carried out have suggested that antisperm antibodies (ASA) can interfere with the fertilizing ability of the spermatozoa. The ASA can act negatively either on the motility of the spermatozoa in the semen, or on their ability to pass through female genital secretions, or on the fusion of the gametes that represent the key events of the fecundation.

In the last years, owing to the improvement and spreading of in vitro fecundation techniques, it has been possible to demonstrate the effect of antibody-bound sperm directly at the level of in vitro gamete interaction.

As far as the semen is concerned, it was demonstrated that the presence of the ASA can reduce the motility and the concentration of the spermatozoa and can induce the formation of spermagglutination. It must be noted that an apparent normozoospermia can mask a presence of a high percentage of ASA bound to the sperm surface or a high ASA titer in serum or seminal plasma [2].

ASA can affect sperm penetration of cervical mucus. In case of ASA present in cervical mucus or bound to the sperm surface we can observe at the PCT an impaired sperm penetration in the cervical mucus or abnormal swimming behavior within the mucus, ranging from complete immobilization of sperm, vibratory motion with limited progression ('shaking phenomenon'), to restricted tail beat frequency and loss of rotatory motion.

The presence of a large percentage of spermatozoa with vibratory motility can also be seen in the semen-cervical mucus contact test (SCMCT). The shaking reaction in these cases is presumably due to the cross-linking of motile, antibody-coated spermatozoa to the cervical mucus gel via the Fc part of the antibody molecule [3]. Various investigations suggested that there was a relationship between sperm-bound autoantibodies of the IgA immunoglobulin class and a poor SCMCT [4] and between sperm antibodies and poor postcoital test results [5]. In a recent work, Clarke [6] confirmed that using the capillary test sperm-bound antibodies of IgA immunoglobulin class are associated with poor in vitro cervical mucus penetration. Analysis of the regional specificities indicates that antibodies to the tail endpiece do not interfere with in vitro cervical mucus penetration while autoantibodies to the sperm tail mainpiece are of primary importance in interfering with in vitro cervical mucus penetration.

Our findings, employing the slide test as the test of sperm-cervical mucus interaction, do not confirm such a strict correlation [7]. In fact, above a cutoff of 60% of positivity, it is really difficult to identify sharp differences in the immunoglobulin class and in the site of the binding that can mainly be related with poor results of the sperm-mucus interaction test. Our results confirm our previous observation [8] regarding the main role of the intensity of the immunization on the other immunological parameters (prevalent site of the binding and prevalent immunoglobulin class).

As we said before, the diffusion of the various techniques of artificial insemination and assisted fecundation allowed us to deepen our knowledge on the effect of the ASA on the fertilizing ability of the spermatozoa. The presence of ASA must be taken into account, in a program of assisted reproduction (AR), according a double point of view, that is to say (I) as a possible complication (ASA production after insemination), and (II) as an interfering factor (ASA capacity to block fetilization process).

Antisperm Antibodies as a Complication of
Assisted Reproductive Techniques

The first aspect refers almost exclusively to intraperitoneal insemination (IPI). With this technique the production of ASA could occur owing to the direct, immediate, intense and 'microtraumatic' contact of the pool of spermatozoa with peritoneum. Our data [9], regarding a trial carried out in collaboration with the 3rd Department of Obstetrics and Gynecology of the University of Milan can be summarized as follows: 20 infertile patients with normal tubal patency were inseminated intraperitoneally with spermatozoa (mean 15 million, range 1–48) prepared by the standard swim-up technique. The occurrence of immunization to spermatozoa was looked for using the gelatin agglutination test (GAT) and the tray agglutination test (TAT). In these methods, an antibody titre up to 1/16 is considered not significant from a clinical point of view. Both tests gave negative results for all the controls (10 pregnant and 10 puerperal patients). Antisperm antibodies were measured before, 20 days and 4–7 months after IPI. In the group of inseminated patients, 18 women with no basal sperm antibodies did not show evidence of antibodies formation after the treatment; in the 2 patients who already had low antibody titer (1/32) there was no increase after insemination. In a recent work, employing the indirect IBT and the sperm immobilizing test for the detection of ASA, it has been confirmed that ASA immunization can take place as a transient phenomenon in subjects who underwent IPI [10]. In conclusion, despite the large number of spermatozoa inseminated and even after several IPI attempts, there was no evidence of 'de novo' production of or increase in already present antisperm antibodies.

*Antisperm Antibodies as an Interfering Factor in
Assisted Reproductive Techniques*

With regard to the possible interferences, ASA can reduce the chances of a successful AR if present either in the man or in the woman [11]. The evidence suggests that ASA may inhibit fertilization by binding specifically to membrane Ag involved in sperm-oocyte interaction [12, 13]. Various authors demonstrated that ASA are able to impair the fertilization process at the level of the acrosome reaction [14] of the zona pellucida recognition and penetration [15], and of the sperm-vitellus interaction [16]. A number of studies have shown that homologous ASA can also interfere with the fertilization of zona-free hamster oocytes by human spermatozoa [17, 18]. However, even if the zona-free hamster oocyte system has been a useful model for initial studies, there is evidence to suggest that this system may be an inadequate approximation to the more complex human fertilization process. First, it has been reported that under some particular conditions sperm that can penetrate zona-free hamster oocytes may not be able to penetrate human zonae [19]. Second, there have been conflicting reports about the degree of correlation between zona-free hamster oocyte assay and routine human IVF [20]. More recently, various studies have been performed using salt-stored human zonae pellucidae. They suggest that sperm auto- and isoantibodies of IgG or IgA immnoglobulin classes can interfere with sperm binding to and penetration of the human zona pellucida. It is not certain, however, whether storage of zonae in high salt solutions may induce alterations in the sperm receptor sites so that they undergo subtle changes in their mechanism of interaction with normal or ASA-coated spermatozoa. In view of the limitations of these fertilization models, it is becoming apparent that more relevant information about the effects of the ASA on human fertilization must be derived from the study of human IVF using viable oocytes. Although the results of the studies in this field are suggestive of a correlation between ASA and impaired fertilizing capacity [21–23]; however, they are difficult to analyze from an epidemiological and statistical point of view. One of the most serious problems is that relatively small numbers of patients with ASA will be found in an individual program, thus making it difficult to compare groups with different levels or types of ASA. Moreover, with small groups of patients, it is also more likely that nonimmunological factors may bias results in a group. For this reason, the most appropriate way to eliminate many of the uncertainties associated with this area of investigation is to employ an experimental model using 'spare' oocytes from consenting patients. This approach, even if it allows us to carry out various experimental programs using whole human oocytes, states a series of ethicolegal problems and is not accepted by all the groups involved in the field.

Antisperm Antibodies in Female Genital Tract
Because of ASA interference with sperm progression in the female genital tract, their presence in the female genital secretion is sufficient to advise against all forms of artificial insemination (AIH, IPI) and direct therapy towards an in vitro fertilization.

Antisperm Antibodies in Female Sera
On the other hand, even the presence of ASA in the serum, constantly associated with positivity in the follicular fluid, means that the serum cannot be used for the culture medium and the oocyte must be washed very

carefully to prevent the contamination of the medium in which the gametes will be coincubated. In collaboration with the FIVET Unit of the 2nd Department of Obstetrics and Gynecology of the University of Rome 'La Sapienza', we found that 5% of the female patients enclosed in a program of assisted reproduction have ASAs at high titre in the serum. So the evaluation of the presence of ASA in the serum of women involved in a program of assisted reproduction seem to be useful to avoid 'unexplained' failure [24].

Antisperm Antibodies Bound to Sperm Surface

Finally, in cases of male immune infertility, the use of various techniques for semen manipulation has been proposed [8]. The percentage of success of these techniques, in terms of effective recovery of spermatozoa not involved in antisperm antibody reaction, are very discouraging so that in a review of therapy for immune infertility, Shulman [25], making reference to the method of sperm washing and insemination, stated that 'this procedure has not had marked success, presumably because of the great difficulty of eluting the antibodies from the sperm cell surface by any method that is also sufficiently gentle to the sperm cell.' In our experience, the preparation of semen for artificial insemination in subjects with antisperm autoantibodies has demonstrated that swim-up is not able to select spermatozoa free of antibodies bound to the sperm surface. All the experiments of migration, dilution, antigenic competition and cryopreservation, carried out to evaluate the theoretical possibility of a better recovery of non-'immunologically compromised' spermatozoa, have confirmed such negative results both with the immunobead test and cytofluorimetric analysis [8].

However, we are currently directing our research towards solving the antisperm antibody problem by looking at the physiological modifications of the sperm membrane which take place during in vitro capacitation of sperm. This approach derives from the idea that during in vivo capacitation in the female genital tract and/or in vitro in capacitating media, there can be a loss of membrane structures that is present at the moment of ejaculation [26]. Some of the structures which undergo these changes are the coat molecules that are acquired by the sperm during epididymal maturation and ejaculation while others are intrinsic to the membrane structure. The hypothesis is, therefore, that some of these coat and membrane molecules could be the antigens against which antibodies are directed. If this is true, by inducing in vitro capacitation, one can obtain not only the elimination of the antibody but also the entire immunocomplex without damaging the sperm membrane. In this way, one can get antibody-free sperm without damaging the plasma membrane. Such sperm are at an advanced stage of maturation, but still useful for in vitro fertilization.

For this purpose we have studied selected post-rise populations of antibody-coated sperm at various time intervals after the onset of in vitro capacitation. Using direct IBT we have obtained some preliminary results indicating a reduction of antisperm antibodies on the sperm surface. Studies of sperm function have demonstrated a maintenance of good motility and structural and functional integrity of the sperm membrane. Our data so far indicate that there is a significant reduction of antisperm antibodies, also in immunological terms. If these results are also confirmed by oocyte fecundation it would offer an excellent human experimental model. In fact, the fecundation of an oocyte by an 'immunologically compromised' sperm, after this capacitating therapy, could confirm that this is one solution to immunological pathologies of the sperm membrane.

In conclusion, the presence of ASA can impair not only the fertilizing ability of the spermatozoa in vivo, but also be a serious factor that prevents the success of the various insemination or fertilization techniques. For this reason a screening of the immunologial factor in the couples who undergo a programme of insemination (AIH and/or IPI) or of fertilization (IVF-ET, GIFT, ZIFT, PROST) is imperative.

References

1   Rumke Ph, Hellinga G: Autoantibodies against spermatozoa in sterile men. Am J Clin Pathol 1959;32:357.
2   Dondero F, Lenzi A, Cerasaro M: Patologia dell'apparato riproduttivo; in Introzzi P (ed): Trattato Italiano di Medicina Interna. Roma, USES Editore, 1987, p 1007.
3   Jager S, Kremer J, Kuiken J, Mulder I: The significance of the Fc part of ASA for the shaking phenomenon in the sperm-cervical mucus contact test. Fertil Steril 1981;36: 792.
4   Jager S, Kremer J, Kuiken J, van Slochteren-Draaisma T: Immunoglobulin class of antispermatozoal antibodies from infertile men and inhibition of in vitro sperm penetration into cervical mucus. Int J Androl 1980;3:1.
5   Kremer J, Jager S, van Slochteren-Draaisma T: The 'unexplained' poor postcoital test. Int J Fertil 1978;23:277.

6 Clarke GN: Immunoglobulin class and regional specificity of antispermatozoal autoantibodies blocking cervical mucus penetration by human spermatozoa. AJRIM 1988;16:135.

7 Lenzi A, Gandini L, Lombardo F, Alfano P, Anticoli-Borsa L, Dondero F: Preliminary data on sperm-mucus interaction and sperm-bound antibodies regarding percent positivity, Ig class and site of the reaction; in Spera G, Gnessi L (eds): Unexplained Infertility: Basic and Clinical Aspects. New York, Raven Press, 1989, p 265.

8 Lenzi A, Gandini L, Claroni F, Lombardo F, Morrone S, Dondero F: Immunological usefulness of semen manipulation for artificial insemination homologous (AIH) in subjects with antisperm antibodies bound to sperm surface. Andrologia 1988;20: 314.

9 Ragni G, Lenzi A, Gandini L, Cristiani C, Lombroso GC, Olivares MD, De Lauretis L, Dondero F, Crosignani PG: Lack of immunization after intraperitoneal insemination (IPI) of spermatozoa. Abstr 46th Ann Meet Am Fertility Soc, October 13–18, 1990.

10 Livi C, Coccia E, Versari L, Pratesi S, Buzzoni P: Does intraperitoneal insemination in absence of prior sensitization carry with it a risk of subsequent immunity to sperm? Fertil Steril 1990;53:137.

11 Clarke GN: Sperm antibodies and human fertilization. AJRIM 1988;17:65.

12 Marquant-Le Guinne B, De Almeida M: Role of guinea-pig sperm autoantigens in capacitation and the acrosome reaction. J Reprod Fertil 1986;77:337.

13 Munoz de Vera G, Marquant-Le Guinne B, De Almeida M, Voisin GA: Role of guinea-pig autoantigens in sperm binding to the zona pellucida and oocyte penetration. J Reprod Fertil 1986;77:347.

14 Srivastava GN, Sheikhnejad RG, Fayrer-Hosken R, Malter H, Brackett BG: Inhibition of fertilization of rabbit ova in vitro by antibody to the inner acrosomal membrane of rabbit spermatozoa. J Exp Zool 1986;238:99.

15 Sailing PM, Lakoski KA: Mouse sperm antigens that partecipate in fertilization. II. Inhibition of sperm penetration through the zona pellucida using monoclonal antibodies. Biol Reprod 1985;33:515.

16 O'Rand MG, Irons GP, Porter GP: Monoclonal antibodies to rabbit sperm autoantigens. I. Inhibition of in vitro fertilization and localization on the egg. Biol Reprod 1984;30:721.

17 Alexander NJ: Antibodies to human spermatozoa impede sperm penetration of cervical mucus or hamster eggs. Fertil Steril 1984;41:433.

18 Haas G, Ausmanas M, Culp L, Tureck RW, Blasco L: The effect of immunoglobulin occurring on human sperm in vivo on the human sperm/hamster ova penetration assay. AJRIM 1985;7:109.

19 Gould JE, Overstreet JW, Yanagimachi H, Yanagimachi R, Katz DF, Hanson FW: What functions of the sperm cell are measured by in vitro fertilization of zona-free hamster eggs? Fertil Steril 1983;40:344.

20 Kuzan FB, Muller CH, Zarutskie PW, Dixan LL, Soulas MR: Human sperm penetration assay as an indication of sperm function in human in vitro fertilization. Fertil Steril 1987;48:282.

21 Junk SM, Matson PL, Yovich JM, Bootsma B, Yovich JL: The fertilization of

human oocytes by spermatozoa from men with antispermatozoal antibodies in semen. J IVF Embryo Transfer 1986;3:350.

22   Mandelbaum SL, Diamond SP, DeCherney AH: Relationship of antisperm antibodies to oocyte fertilization in in vitro fertilization-embryo transfer. Fertil Steril 1987;47:644.

23   Mathur S, Mathur RS, Holtz GL, Rust PF, Williamson HO: Cytotoxic sperm antibodies and in vitro fertilization of mature oocytes: A preliminary report. J IVF Embryo Transfer 1987;4:177.

24   Micara G, Moro M, Aragona C, Gandini L, Lenzi A, Dondero F: Evaluation of the male factor in relation to in vitro fertilization; in Capitanio G (ed): GIFT – From Basic to Clinics. New York, Raven Press, p 432.

25   Shulman S: Therapy of immunological infertility. EOS 1986;6:95.

26   Hinrichsen-Kohane AC, Hinrichsen MJ, Schill WB: Molecular events leading to fertilization. A review. Andrologia 1984;16:321.

Dr. Franco Dondero, Laboratory of Seminology and Immunology of Reproduction, V Medical Clinic, Faculty of Medicine, Policlinico 'Umberto I', University of Rome 'La Sapienza', I–00185 Rome (Italy)

Colpi GM, Pozza D (eds): Diagnosing Male Infertility.
Prog Reprod Biol Med. Basel, Karger, 1992, vol 15, pp 186–193

Determination of the Clinical Significance of Antisperm Antibodies Using an in vivo Insemination Technique

Harriet G. Adelson, Jerome H. Check, Jeffrey Chase, Marie Press

The UMDNJ Robert Wood Johnson Medical School at Camden,
Cooper Hospital/University Medical Center, Department OB/GYN,
Division of Reproductive Endocrinology and Infertility, Camden, N.J., USA

The immunobead assay has the advantage of being able to directly determine the presence of antisperm antibodies (ASA) on the sperm surface [1, 2]. However, it is a qualitative rather than quantitative test; the test determines the percentage of the sperm coated by ASA but the assay does not provide any idea of how many antibodies are attached to each sperm.

The intrinsic quality of the sperm, motile density, and concentration of the antibody on the sperm surface may all effect the clinical significance of the presence of antibodies as measured by the immunobead assay. In fact, the present cut-off level of 50% that distinguishes normal from subnormal was arbitrarily chosen based on the fact that many normal donors have levels up to that point [3].

A highly effective therapy for cervical factor has been previously described using a dose of ethinyl estradiol that would suppress ovulation but maximally stimulate cervical mucus; follicular maturation would be simultaneously stimulated with human menopausal gonadotropins [4, 5]. It was reasoned that by using the high dose estrogen to maximally stimulate mucus and block ovulation one could determine the likelihood of an immunological problem by the comparison of first male partner's sperm and then donor sperm following therapeutic donor insemination (TDI). Theoretically, demonstration of sperm with progressive forward motion (PFM) after insemination of male partner's sperm (AIH) would suggest absence of ASA in semen or mucus; failure to show PFM following AIH but showing PFM following TDI would suggest ASA in the male partner; failure to demonstrate PFM following AIH and TDI would suggest ASA in cervical mucus.

A study was thus performed to see the correlation of ASA as detected by direct immunobead test (IBT) on the sperm and by indirect IBT on the mucus with the PFM of sperm following AIH and TDI. Furthermore, in patients where ASA was suspected as being etiologic the couple would try intracervical inseminations for 6 months without hyperstimulation. Those failing to conceive would be offered corticosteroids and the pregnancy rates for the next 6 months were then determined.

Finally, the presence of positive ASA may render the sperm incapable of fertilization without necessarily causing a poor postcoital test. For example, antibodies could prevent sperm attachment to the zona pellucida, therefore to evaluate the pathologic effect of ASA even when the postcoital test was adequate consecutive couples composed of positive and negative patients with good and poor postcoital tests would have their pregnancy rates compared over a 6-month time period.

Materials and Methods

Study 1: Unexplained Poor Postcoital Tests

A total of 30 couples were selected where suspicion of a possible immunological etiology for a poor postcoital test was based on demonstration of normal quality mucus with a Moghissi score above 10 [6]. The semen analysis was required to have a minimum volume of 1.5 ml, 20×10^6 sperm/ml with a minimum of 60% with PFM. Each patient was required to demonstrate at least five sperm/high-powered field but with none demonstrating PFM. Furthermore, no other infertility factors were apparent except for the poor post-coital test.

A direct IBT was performed on all semen specimens using the technique described by Bronson et al. [2]. The percent of IgA and IgG were measured. An indirect immunobead assay was performed on cervical mucus which was heat inactivated to remove complement. Antibody-free donor sperm were employed; collection was obtained by a 'swim-up' technique [7]. An aliquot of antibody-free sperm was then incubated with the cervical mucus to allow for the antibody attachment to the sperm. The direct IBT was performed on these donor sperm specimens that had been exposed to cervical mucus.

High-dose estrogen (ethinyl estradiol, 50 µg daily) is given from day 3 to day 14 of cycle. On day 14 a serum estradiol (E_2) is obtained to ensure breakthrough ovulation did not occur (ethinyl estradiol does not cross-react with the 17 Beta E_2 assay). The patient is then ready for AIH and assessment of sperm movement in the mucus which is performed at 2 and 8 h. Poor PFM leads to repeat testing the next day after TDI.

Study 2: Evaluation of Consecutive Patients with a Female Infertility Factor Identified (Other than Cervical Factor) and Corrected – Influence of Antisperm Antibodies in the Male on Postcoital Testing and Pregnancy Rates following Timed Intrauterine Insemination

Another group, consisting of 60 consecutive patients were evaluated to compare the correlation of postcoital testing with ASA. A good-to-fair postcoital test was considered if at least 3–5 sperm/high-powered field with PFM were seen at least 8 h after intercourse. Pregnancy rates after 6 months were then determined. Those couples with poor postcoital tests were treated with timed intrauterine insemination (IUI) (approximately 36 h from initiation of serum luteinizing hormone (LH) surge or 12 h from peak LH surge; daily LH levels were obtained beginning 10 days prior to expected menses until ovulation was determined).

Results

Three groups were separated based on the response of the female partners to AIH and TDI. Group 1 consisted of 8 couples who demonstrated sperm with PFM following AIH. This group would not be expected to have either ASA in the semen or cervical mucus. Group 2 was composed of 17 couples where no sperm with PFM were seen following AIH but PFM was seen following TDI; this group would be predicted to have the largest percentage of male partners with positive ASA levels. The final 5 patients comprised group 3 where neither AIH nor TDI resulted in sperm with PFM. Cervical mucus ASA would be predicted.

The incidence of positive ASA ($\geq 50\%$) as determined by IBT according to different groups is seen in table 1. Slightly positive was considered between 20 and 49%. The response to high-dose estrogen did in fact predict the groups with the significant ASA levels. Not one of the Group 1 patients demonstrated positive ASA whereas 47% of group 2 males were positive and if slightly positive is counted, the level was 71%. In contrast, not one of the females were positive for ASA. Group 3 was in fact the only one demonstrating ASA in cervical mucus (40%).

The differential response of AIH and TDI to the estrogen stimulated mucus and the presence or absence of ASA determined the type of therapy. Group 1 women were treated with the high dose estrogen-hMG technique using ethinyl estradiol [5] and 4/8 (50%) conceived in 6 months; 2/4 (50%) of the remainders conceived during the next 6 months with the same technique and timed IUI. A total of 3/17 (17.6%) of group 2 females became pregnant following intracervical insemination. These pregnancies were

Table 1. Incidence of positive ASA in groups 1–3

	Total	+IgG	+IgA	+IgG or IgA	Percent +ASA	PCT + ASA IgG or IgA	+ASA and PCT+
Group 1							
Male	8	0	0	0	0	0	0
Female	8	0	0	0	0	0	0
Group 2							
Male	17	8	4	8	47	4	12 (70.6%)
Female	0	0	0	0	0	0	0
Group 3							
Male	5	1	0	0	20	1	2 (40%)
Female	5	2	2	2	40	0	2 (40%)

achieved in the female partners of the males without ASA (3/9, 33%). None of the 8 females with partners with positive ASA became pregnant during the first 6 months following intracervical insemination but 6/8 (75%) conceived following male treatment with methylprednisolone 96 mg × 1 week each month during the next 6 months. One of 2 couples with slightly positive ASA in the males treated with corticosteroids also achieved pregnancies during the second 6 months. All couples conceiving following methylprednisolone therapy first demonstrated sperm with PFM in the mucus. One of 5 group 3 females conceived following intracervical insemination during the first 6 months; 1 of 2 having positive ASA in the mucus was treated with corticosteroids during the second 6 months and she conceived.

The correlation of postcoital tests and presence of antisperm antibodies and pregnancy rates (6 months) following timed intrauterine insemination is seen in table 2. One of the original 60 patients dropped out of the study. Only 9% of couples (4/44) with normal postcoital tests were positive for ASA compared to 60% of couples with poor postcoital tests (9/15). Statistical analysis was performed using the Fisher's exact test; p < 0.001. There was a higher pregnancy rate over 6 months in the group with normal postcoital tests (84%) compared to those with poor results treated by timed IUI (67%). However the Fisher's exact test did not show statistical differ-

Table 2. Correlation of postcoital tests and presence of ASA and pregnancy rates (6 months)

	Total	Pregnant		+ASA		Pregnancy +ASA		−ASA	Pregnancy −ASA	
		n	%	n	%	n	%		n	%
Good postcoital tests	44	37	84	4	9	4	100	40	33	83
Poor postcoital tests	15	10	67	9	10	5	56	6	5	83

+ASA refers to a level over 50% employing immunobead test on the spermatozoa.

ence, p = 0.14. Considering all patients with positive ASA only 4/13 (30%) had normal postcoital tests compared to 40/46 patients (87%) with negative ASA. Fisher's exact test was significant, p < 0.001. The pregnancy rate in the couples with positive ASA in the male partner was 69% (9/13) compared to 83% (38/46) with negative ASA in the male partner but the results were not statistically different (Fisher's exact test p = 0.24).

The cumulative probability of pregnancy in couples with positive ASA is seen in table 3a, negative ASA in table 3b, good postcoital tests in table 3c and poor postcoital tests in table 3d.

Discussion

The data from the first study demonstrated that not all patients with unexplained poor postcoital tests have an immunological etiology. The differential response to AIH or TDI in estrogen-stimulated mucus was able to predict fairly well whether ASA is present or not and in which partner. Though only 47% of group 2 men were found positive for ASA the slightly positive levels ≥20% but <50% may still have contributed so that if positive is considered ≥20% then 71% would be positive. Thus, where reliable immunobead assays are not available, a reasonable idea of whether ASA is present may be obtained by this insemination technique. Furthermore, the data showed that the majority of couples with unexplained poor postcoital tests do not necessarily have an immunological basis (only 33%). Although the Moghissi score was over 10 in all cases, it is of interest that in 27% (8/30) of the cases improving the mucus even further with estrogen

Table 3. Cumulative probability of pregnancy in couples under different conditions

Number of cycles	Number of patients	Pregnancies achieved	Lost to follow-up	Patient months of treatment	Pregnancy rate per month	Cumulative probability of pregnancy
a Positive ASA						
1	13	6	0	10.00	0.60	0.60
2	7	1	0	6.50	0.15	0.66
3	6	0	0	6.00	0.00	0.66
4	6	0	0	6.00	0.00	0.66
5	6	0	0	6.00	0.00	0.66
6	6	2	0	5.00	0.40	0.80
		9				
b Negative ASA						
1	46	17	0	37.50	0.45	0.45
2	29	4	0	27.00	0.15	0.53
3	25	3	0	23.50	0.13	0.59
4	22	8	0	18.00	0.44	0.77
5	14	2	0	13.00	0.15	0.81
6	12	4	0	10.00	0.40	0.89
		38				
c Good postcoital						
1	44	17	0	35.50	0.48	0.48
2	27	3	0	25.50	0.12	0.54
3	24	2	0	23.00	0.09	0.58
4	22	7	0	18.50	0.38	0.74
5	15	2	0	14.00	0.14	0.78
6	13	6	0	10.00	0.60	0.91
		37				
d Poor postcoital						
1	15	6	0	12.00	0.50	0.50
2	9	2	0	8.00	0.25	0.63
3	7	1	0	6.50	0.15	0.68
4	6	1	0	5.50	0.18	0.74
5	5	0	0	5.00	0.00	0.74
6	5	0	0	5.00	0.00	0.74
		10				

can result in improved postcoital tests. Thus, if the estrogen alone allows PFM with AIH one can use the ethinyl estradiol-hMG technique to try for pregnancy.

The second study did not require good quality cervical mucus and yet the positive ASA in the male was still present in 60% (9/15) of the cases. The fact that all 4 of the couples with positive ASA in the males with good to fair postcoital tests conceived, does not support the concept that even if sperm progression in the mucus is not impaired, probably some other aspect of the fertilization process is adversely effected. Thus, one should never treat a patient merely on the presence of ASA but must first demonstrate a way that the immunological factor is causing infertility. It is clear, however, that only a minority of patients with good postcoital tests (9%) are positive for ASA whereas a majority (60%) are negative for ASA.

Achievement of good postcoital tests resulted in a better pregnancy rate (84%) than timed IUI by LH surge (67%) but no statistical difference was seen. However, the fact that a reasonable pregnancy rate of 67% over 6 months was accomplished by timed IUI, the latter seems to be a more appropriate therapy than corticosteroid therapy (which resulted in a 75% pregnancy (6/8) rate in the partner of the treated males), since one of the side effects of therapy could be aseptic necrosis of the femoral heads [8, 9].

Many clinicians when treating with IUI also hyperstimulate the patient with hMG, which increases the cost and exposes the patient to a greater risk of multiple gestation and ovarian hyperstimulation syndrome. Perhaps by carefully observing the serum LH levels a good 6-month pregnancy rate can be achieved without the use of hMG.

References

1   Clarke GN, Stojanoff A, Cauchi MN, McBain JC, Speirs AL, Johnston WI: Detection of antispermatozoal antibodies of IgA class in cervical mucus. Am J Reprod Immunol 1984;5:61–65.
2   Bronson R, Cooper G, Rosenfeld D: Ability of antibody-bound human sperm to penetrate zona-free hamster ova in vitro. Fertil Steril 1981;36:778–783.
3   Jennings MG, McGowan MP, Baker HWG: Immunoglobulins of human sperm: Validation of a screening test for sperm autoimmunity. Clin Reprod Fertil 1985;3: 335–342.
4   Check JH, Adelson HG: Improvement of cervical factor by high-dose estrogen and human menopausal gonadotropin therapy with ultrasound monitoring. Obstet Gynecol 1984;63:179–181.

5 Check JH, Wu CH, Dietterich C, Lauer CC, Liss J: The treatment of cervical factor with ethinyl estradiol and human menopausal gonadotropins. Int J Fertil 1986;31: 148–152.

6 Moghissi KS: The cervix in infertility. Clin Obstet Gynecol 1979;22:27–42.

7 Check JH, Shanis BS, Cooper SO, Bollendorf A: Male sex preselection: Swim-up technique and insemination of women after ovulation induction. Arch Androl 23: 101–102.

8 Solomon L: Drug-induced arthropathy and necrosis of the femoral head. J Bone Joint Surg 1983;55B:246–261.

9 Parker LN: Corticosteroid therapy and aseptic necrosis. Ann Intern Med 1983;9: 882.

Jerome H. Check, MD, 7447 Old York Road, Melrose Park, PA 19126 (USA)

Colpi GM, Pozza D (eds): Diagnosing Male Infertility.
Prog Reprod Biol Med. Basel, Karger, 1992, vol 15, pp 194–201

A Quantitative Study of Ultrastructural Spermatozoa Defects in Relation with Infertility

D. Raick, E. Baeckeland, J. Letawe

Laboratory of Embryology, University of Liège, Belgium

Since the studies of Mc Leod and Gold [1], measuring the differences in sperm characteristics between fertile and infertile males, many authors have shown that such parameters as concentration, motility and morphology are on average more superior in fertile than in infertile males. Nevertheless, the respective role of various sperm characteristics is still controversial. Whereas some authors assert for instance that sperm morphology may not be important in predicting fertility in subfertile men with a mean sperm concentration of 5 million/ml [2], others have shown that the percentage of sperm with normal morphology was significantly higher when a pregnancy occurred than when the couple remained infertile [3]. The human in vitro fertilization has greatly improved the understanding of the significance of this parameter for fertilization. In this field, Kruger et al. [4, 5] have shown that by evaluating sperm morphology with strict criteria, the percentage of normal sperm has an important role in the fertilization and a high prognostic value. They indicate that there is a clear threshold in normal sperm morphologic feature at 14% with high fertilization and pregnancy rates in the group with normal sperm morphologic feature > 14%. Jeulin et al. [6] point out that the morphology of the acrosome is of primordial importance to obtain a success in in vitro fertilization. This observation seems to be in contradiction with studies on acrosome reaction that failed to demonstrate a correlation between the percentage of reacted sperm and penetration capacity [7–9]. For the most part, this apparent contradiction could be due to the fact that the status of this particular sperm organelle cannot be evaluated precisely by conventional methods of semen analysis. Similarly, Zamboni [10] draws attention on the existence and makes a review of a variety of sperm defects that are causes of infertility and are not identifiable by photonic microscopy. He advances the potential and relevance of the ultrastructural examination of the semen for the diagnostic assessment of infertile conditions and for their prognosis and management.

Because of the well-known pleiomorphism of the human sperm still pronounced by transmission electron microscopy and of the fact that ultrastructural abnormalities may be found even in spermatozoa of normal and fertile individuals, the diagnosis of a true pathologic situation needs a quantification of these abnormalities, that is to say a careful examination of a large number of spermatozoa studied on plane of section favorable to the visualization of the relevant structures. This requirement makes the ultrastructural assessment of semen heavy and time consuming. It is perhaps the reason why, in spite of an extraordinary potential, this technique has a limited application.

Until now, only two research teams have adopted this process [11, 12]. In the present article, we expose our own method of ultrastructural quantitative examination of sperm and we compare the results obtained on 19 fertile and 162 suspected infertile men.

Material and Methods

Subjects

The semen samples were obtained from two groups of men:

(1) *Control group:* this consisted of 19 fertile men (18–35 years old) having fathered 1 child within the last year before the analysis of their semen, or 2 children within the last 3 years. They had the following semen characteristics: > 20 million spermatozoa/ml; > 60 % motile spermatozoa; > 40 % normal forms.

(2) *Subfertile group:* this consisted of 162 men attending an andrologic consultation after a period of infertility longer than 1 year. Patients presenting a pathological state related to infertility as urogenital infection, cryptorchidism or varicocele were excluded from this study. Within this group, 47 patients had semen characteristics that could be considered as normal: > 20 million spermatozoa/ ml; > 60 % motile spermatozoa; > 60 % normal forms. For the 115 other men, one of the semen parameters at least was below these values.

Preparation of Samples

All semen samples were collected at the laboratory, by masturbation. After liquefaction (max. 30 min), 1 ml of sperm was centrifuged at 4,500 rpm for 10 min. The pellet was cut into small pieces, fixed for 30 min with 2.5 % glutaraldehyde in 0.1 M cacodylate buffer at 4 °C, washed 3 times with that buffer, postfixed for 30 min in 1 % OsO_4 in 0.1 M cacodylate buffer at 4 °C and then washed 3 times with bidistilled water. The specimens were dehydrated through a series of ascending concentrations (70, 95 and 3 times 100%) of ethanol and with propylene oxide (2 times for 10 min). During the dehydratation, the specimens were brought back to room temperature. Samples embedded in Epon 812 were sectioned with a Reichert OMU 3 ultramicrotome, contrasted with uranyl acetate for 15 min followed by lead citrate for 15 min at room temperature [13]. Preparations were examined in a Philips EM 201 electron microscope.

Photonic Analysis

During the fixation of the samples for electron microscopy, a routine seminal analysis according to standard procedures was performed. Concentration and motility of the spermatozoa were evaluated in a Makler chamber at a 160 × magnification. The percentage of living cells was established from a smear after incubation of a drop of semen with eosin yellow and nigrosin [14]. For morphology analysis, 100 spermatozoa on a stained smear (Feulgen-Light Green) were analyzed at a 1,000 × magnification according to the classification of David et al. [15].

Ultrastructural Quantitative Analysis

The ultrastructural examination was performed on at least 60 longitudinal sections of heads passing through the connecting piece and at least 60 transverse sections of tails (1/3 through the midpiece and 2/3 through the principal piece – sections through the distal piece of the tail were excluded since at this level, the longitudinal columns, dense fibers and some dynein arms are absent). The quantification of the abnormalities was realized only on sections with a well-preserved plasma membrane to exclude the possible artifactual degradations of the organelles. Spermatozoa showing signs of necrosis and lysis were considered separately.

Concerning the heads, we distinguished 6 groups of abnormalities: (a) 3 for the nucleus: presence of large intranuclear vacuoles; immaturity of the chromatin; presence of two or more nuclei; (b) 2 for the acrosome: abnormal acrosome (this group included abnormalities of structure and abnormalities of position of the acrosome); premature acrosome reactions; (c) 1 for the cytoplasm: presence of residual cytoplasm.

Concerning the axonemes, we distinguished 9 groups of abnormalities: (1) absence of the inner and outer dynein arms; (2) Absence of the inner dynein arms and peripheral junctions; (3) absence of the outer dynein arms; (4) absence of the radial spokes; (5) absence of the central complex; (6) absence of one or more doublets; (7) presence of supernumerary doublets; (8) total disorganization of the axoneme; (9) presence of two or more axonemes.

Since the abnormalities at the level of the head as well as the level of the axoneme could be isolated or associated, we used a multiple entry system (inspirated of those proposed by David et al. [15] for photonic analysis) to perform our analysis with electron microscope. The results of the quantitative ultrastructural study are expressed, for the abnormalities of the head, as percentages of longitudinal sections of heads in good state (no sign of lysis) and for the abnormalities of the axoneme, as percentages of transverse sections of tails in good state.

Statistical Analysis

In order to compare the fertile and subfertile groups, the statistical methods used were simple tests: t test and analysis of variance (ANOVA).

Table 1. Sperm characteristics measured in the fertile and subfertile groups

	Group I: fertile men	Group II: subfertile men
Number of men	19	162
Age	28.6 (3.9)	30.2 (4.9)
Volume, ml	3.0 (1.4)	3.1 (1.5)
Concentration, 10^6/ml	135.8 (59.9)	81.9 (86.7)**
Motility, %	76.9 (11.0)	59.6 (20.6)**
Morphologically abnormal spermatozoa, %	36.5 (12.5)	45.1 (18.9)*
Round cells, 10^6/ml	2.9 (4.5)	1.9 (3.8)
Necrospermia, %	14.6 (8.2)	19.8 (9.9)*

SD values are in parentheses.
* $p < 0.05$, ** $p < 0.005$, in comparison between the fertile and subfertile group.

Results

Photonic Microscopy

Mean values for semen characteristics in the fertile and subfertile groups are shown in table 1. We observe a significant difference between the two groups concerning the mean concentration, motility, percentage of abnormal forms and necrospermia. The highest statistical differences are found for concentration and motility, but the mean characteristic values obtained in the two groups are close to the values usually considered as normal.

Electron Microscopy

Using our own method of quantitative ultrastructural examination of semen we have measured the mean percentages of ultrastructural abnormalities in the fertile and subfertile groups. The results are shown in table 2. On the basis of the criteria described in the Material and Methods section, we have established reference values from our 19 fertile donors. We observe that there are, in semen considered as normal, 41.5% (SD $\pm$ 9.8%) abnormal heads and 25.7% (SD $\pm$ 9.8%) abnormal axonema. The abnormalities more commonly encountered are the presence of an abnormal acrosome (18.5%, SD $\pm$ 6.2%), the presence of residual cytoplasm (16.2%, SD $\pm$ 7.2%) and the nuclei presenting an immature chromatin

Table 2. Pecentages of ultrastructural abnormalities of the spermatozoa measured in the fertile and subfertile groups

	Group I: fertile men	Group II: subfertile men
Lysis	12.4 (8.5)	18.6 (13.8)*
Abnormal heads	45.1 (9.8)	62.6 (19.3)**
Intranuclear vacuole	6.7 (4.9)	8.7 (7.1)
Immature chromatin	11.2 (8.8)	21.2 (20.8)*
Two or more nuclei	3.8 (3.9)	4.5 (6.8)
Abnormal acrosome	18.5 (6.2)	36.3 (17.5)**
Premature acrosome reaction	8.1 (4.4)	6.7 (9.5)
Residual cytoplasm	16.2 (7.2)	27.5 (19.3)
Abnormal axonema	25.7 (9.8)	41.9 (19.7)**
Absence of inner and outer dynein arms	0.1 (0.4)	0.1 (0.8)
Absence of inner dynein arms and peripheral junctions	0	0.4 (4.0)
Absence of outer dynein arms	0	0.4 (2.1)
Absence of radial spokes	0.3 (0.7)	1.0 (2.1)
Absence of central complex	5.0 (4.8)	6.3 (7.2)
Absence of one or more doublets	16.1 (8.8)	25.6 (14.5)**
Supernumerary doublet(s)	1.8 (1.6)	1.4 (2.6)
Total disorganization	3.7 (3.3)	6.5 (9.7)
Two or more axonema	0.5 (0.8)	3.3 (4.7)
Abnormal periaxonemal structures	0	2.2 (10.2)

SD values are in parentheses.
* $p < 0.05$, ** $p < 0.005$, in comparison between the fertile and subfertile group.

(11.2%, SD $\pm$ 8.8%). The frequencies of the other defects considered are below 10%. In the fertile group the percentage of abnormal acrosomes never exceeds 35%. Concerning the axonemes, the only abnormality exceeding 10% is the absence of one or more doublets (16.1%, SD $\pm$ 8.8%).

In the infertile group the mean percentage of abnormal heads rises to 62.6% (SD $\pm$ 19.3%) and those of abnormal axonema to 41.9% (SD $\pm$ 19.7%). The most frequent abnormalities are the same as in the fertile group but their percentages are higher. The mean percentage of abnormal acrosomes in the subfertile group is 36.3% (SD $\pm$ 17.5%).

Comparison of Fertile and Infertile Groups

Among the 19 categories of abnormalities we distinguished, 18 show a higher percentage in the subfertile group. The frequency of premature acrosome reaction is the only parameter that is somewhat lower in the subfertile group but this difference is not significant. We observe a significant difference between the two groups concerning the lysis, the global percentage of abnormal heads and the global percentage of abnormal axonema. If we consider the detailed abnormalities of the heads, two show significant difference with higher percentages in the infertile group: the presence of immature chromatin and the abnormalities of the acrosome; concerning the axonemes, the only abnormality that presents a significant higher frequency in the infertile group is the absence of one or more doublets.

The most discriminant parameters (p < 0.005) between the two groups seem to be: (1) the global percentages of abnormal heads and axonemes; (2) the percentage of abnormal acrosome, and (3) the percentage of axonema lacking one or more doublets.

Discussion

As in most of the studies, our results of conventional semen analysis show that the mean sperm concentration, motility, percentages of normal and vital cells are lower in a population of infertile men than in fathers who did not consult for infertility. The fact that in our study, the mean semen characteristics are so 'good' can easily be explained by the fact that this group is composed of selected subjects. Men with azoospermia or severe oligozoospermia are excluded for technical reasons. Men with severe pathological states known to be associated with infertility (urogenital infection, varicocele, cryptorchidism) are also excluded because we chose to study healthy men with no severe disturbance of the semen parameters and who, however, cannot conceive. In other words, the subfertile group in question is composed of 'limit' cases sent to us by clinicians in order to settle if they must be considered as pathological or as normal. The results of the conventional analysis presented here reflect this selection.

In the field of male infertility, the pleiomorphism of the spermatozoa makes it very difficult to establish limits of normality. What is true for conventional semen parameters is even more true for ultrastructural parameters. This difficulty is linked to the very large distributions of the percentages of ultrastructural abnormalities observed in the subfertile group (see the SD) and to the overlapping existing between fertile and sub-fertile groups. However and although our subfertile group is composed of 'limit' cases, we observe that the global percentages of abnormal heads and axonemes measured in this group are markedly and significantly higher than the reference values obtained from the fertile group. We conclude that there is in the subfertile subjects an ultrastructural pathology of the spermatozoa. This pathology can affect the heads, axonemes, or both.

The statistical analysis shows that two abnormalities seem to be highly discriminant between the two groups: the first one affects the head and consists in acrosome defects. The mean percentage of abnormal acrosome measured in the subfertile group is twice the mean percentage obtained in the fertile one and is higher than the maximum value observed for this parameter in the fertile subjects (34%). This result is very important; it confirms Jeulin's results about the relation between acrosomal morphology and ability to fertilize. The second discriminant defect affects the axoneme and consists in the absence of one or more doublets. This abnormality is the most frequently encountered in the fertile as in the subfertile group; this observation is in agreement with the results of Escalier et al. [16] who had described and quantified the axonemal abnormalities of spermatozoa obtained from 56 infertile patients.

The last but less significant difference between the two groups concerns the percentage of heads with an immature chromatin. It has been shown that this particular defect of the nucleus could be in itself capable of preventing fertility.

In conclusion, the comparative quantitative study of ultrastructural abnormalities of the spermatozoa in a fertile and a subfertile group shows that: (1) the nature and the relative frequencies of the abnormalities are the same in the two groups; (2) the percentages of ultrastructural abnormalities are higher in the spermatozoa of subfertile subjects, and (3) the more discriminant abnormalities between the two groups are the defects of the acrosome, the absence of one or more doublets at the level of the axonemal complex and the immaturity of the chromatin.

These results are in close agreement with the knowledge one has about the implication of those different components of the spermatozoon in the processus of fertilization. On the basis of the data presented here, we are not yet able to establish a definitive clear-cut limit of normality for each of these parameters. The ultrastructural study of sperms that fail to fertilize eggs in in vitro fertilization might help us to fix limits above which fecundity is almost null. This will be the next stage of our research.

References

1   McLeod J, Gold RZ: The male factor in fertility and infertility. IV. Sperm morphology in fertile and infertile marriage. Fertil Steril 1951;2:394–414.

2   Zaini A, Jennings HG, Backer HGW: Are conventional sperm morphology and motility assessments of predictive value in subfertile men? Int J Androl 1985;8: 427–435.

3   Jouannet P, Ducot B, Feneux D, Spira A: Male factors and the likelihood of pregnancy in infertile couples. I. Study of sperm characteristics. Int J Androl 1988;11: 379–394.

4   Kruger TF, Menkveld R, Stander FSH, Lombard CJ, Van der Merwe JP, Van Zyl JA, Smith K: Sperm morphologic features as a prognostic factor in in vitro fertilization. Fertil Steril 1986;46:1118–1123.

5   Kruger TF, Acosta AA, Simmons KF, Swanson RF, Matta JF, Oehninger S: Predictive value of abnormal sperm morphology in in vitro fertilization. Fertil Steril 1988; 49:112–117.

6   Jeulin C, Feneux D, Serres C, Jouannet P, Guillet-Rosso F, Belaisch-Allart J, Frydman R, Testart J: Sperm factors related to failure of human in vitro fertilization. J Reprod Fertil 1986;76:735–744.

7   Plachot M, Mandelbaum J, Junca A-M: Moyens actuels d'évaluation de la fécondance du sperme. Contrib Fertil Sex 1986;14:211–219.

8   Yang YS, Rojas FJ, Stone SC: Acrosome reaction of human spermatozoa in zona-free hamster egg penetration test. Fertil Steril 1988;50:954–959.

9   Fukuda M, Cross NL, Cummings-Paulson L, Yee B: Correlation of acrosomal status and sperm performance in the sperm penetration assay. Fertil Steril 1989;52:836–841.

10 Zamboni L: The ultrastructural pathology of the spermatozoon as a cause of infertility: the role of electron microscopy in the evaluation of semen quality. Fertil Steril 1987;48:711–734.

11 Bartoov B, Eltes F, Langsam J, Snyder J, Fisher J: Ultrastructural studies in morphological assessment of human spermatozoa. Int J Androl 1982;5(suppl):81–96.

12 Escalier D, Bisson JP: Quantitative ultrastructural modifications in human spermatozoa after freezing; in David G, Price WS (eds): Human Artificial Insemination and Semen Preservation. Plenum, New York, 1979, pp 107–122.

13 Reynolds E: The use of lead citrate at high pH as an electron opaque stain in electron microscopy. J Cell Biol 1963;17:208–212.

14 Eliasson R: Supravital staining of human spermatozoa. Fertil Steril 1977;28:1257.

15 David G, Bisson JP, Czyglik F, Jouannet P, Gernigon C: Anomalies morphologiques du spermatozoïde humain. 1. Propositions pour un système de classification. J Gynecol Obstet Biol Reprod 1975;4:17–36.

16 Escalier D, David G: Pathology of the cytoskeleton of the human sperm flagellum: Axonemal and periaxonemal anomalies. Biol Cell 1984;50:37–52.

Dr. D. Raick, Laboratory of Embryology, University of Liège, 20, rue de Pitteurs, B-4020 Liège (Belgium)

Colpi GM, Pozza D (eds): Diagnosing Male Infertility.
Prog Reprod Biol Med. Basel, Karger, 1992, vol 15, pp 202–209

Fertility of OAT Patients Assessed by the Results of Assisted Reproductive Technology

C. Cimino, E. Cittadini, F. Gattuccio

Istituto Materno-Infantile, Università di Palermo, Italy

A retrospective look at our IVF/ET program reveals that male subfertility is becoming more and more frequent as an indication for the use of this technique. Up till December 1985 this problem involved only 6% of the cases treated, whereas in 1986 and the first trimester of 1987 it occurred in 80 of 324 cases (24.7%) (fig. 1). The tendency to use assisted conception techniques in such couples is due to the disappointing results often obtained with the use of classical andrological therapy, either direct or not, in cases of dyspermia, particularly the idiopathic type.

Our own use of assisted fecundation techniques in male subfertility has for a long time been based on the attempt to bring the pregnancy rate in dyspermic couples up to the same level as that obtained in normospermic cases, by means of careful preliminary evaluation of the seminal fluid, and it is from this point of view that the use of the pellet swim-up test (PST) should be seen [1]. PST adds another parameter to traditional methods of seminal fluid assessment which must be performed before deciding to include the patient in an assisted fecundation program (table 1).

The results of the PST make it possible to divide OAT patients into two groups: one with 1.5×10^6/ml spermatozoa or more, and the other with less than this. Normospermic patients generally give much higher PST results than this. With traditional methods of seminal fluid assessment we have considered as subfertile those patients who showed less than 12×10^6/ml spermatozoa with forward progressive and rapidly forward progressive motility 60 min after ejaculation; men with this value or higher have been considered as normospermic.

Table 2 shows several selected series of assisted conception cycles performed during the 3 years from 1987 to 1989 with the pregnancy rate

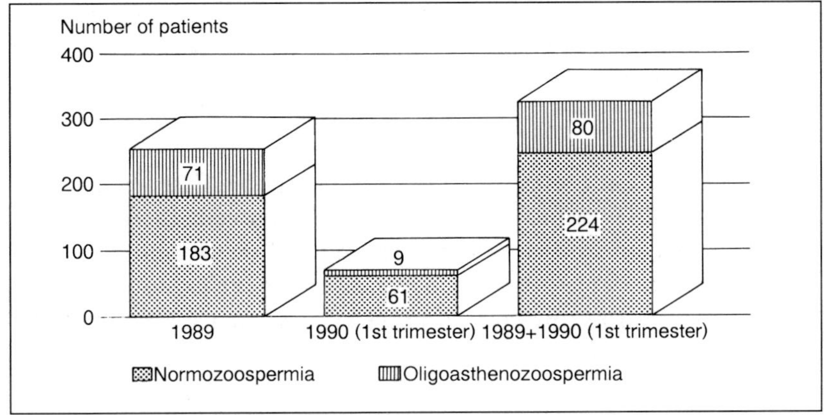

Fig. 1. IVF programme at the University of Palermo. Male subfertility indications.

Table 1. Pellet swim-up test

Centrifugation at 1,800 rpm of 0.5 ml of sperm with 1.5 ml of Ham's F 10 for 10 min
Elimination of the supernatant, layering of 0.5 ml of Ham's F 10 on the pellet and incubation at 35 °C for 2 h

Table 2. Selected series of assisted conception cycles (1987–1989)

		Patients	Pregnancies	%
IVF/ET	NS	244	35	14.3
	DS (PST+)	80	9	11.2
GIFT	NS	318	112	35.2
	DS (PST+)	152	30	19.7
ZIFT[1]	DS (PST–)	22	2	9.1
DIPI	NS	122	39	31.9
	DS (PST+)	115	24	20.8
	DS (PST–)	105	6	5.7

[1] ZIFT performed: 10; pregnancy rate: 20%.
NS = Normospermic; DS = dyspermic; PST = pellet swim-up test.

assessed according to the presence of male subfertility. In this series of IVF/ET [2], performed between January 1986 and March 1987, positive PST enabled us to select a group of dyspermic patients and obtain a pregnancy rate per patient of 11.2% which was not significantly different from that obtained in normospermic patients.

The GIFT data refer to the year 1988 and the first semester of 1989 [3]. Positive PST results do not necessarily prevent the pregnancy rate from being significantly lower in dyspermic couples from that obtained in normospermic couples; they do, however, lead to an extremely satisfactory pregnancy rate from the clinical point of view, at 19.7% per patient and 16.9% per GIFT cycle.

With ZIFT [4], performed during the second trimester of 1989, we obtained encouraging results, bearing in mind the fact that all the patients undergoing this technique showed negative PST and that the pregnancy rate (20%) refers to the actual number of ZIFT's performed and not to the rate per patient (9.1%) which also includes those cases where no oocyte was fertilized, and no transcelioscopic intratubal transfer of zygotes was therefore possible.

We obtained fairly good results in dyspermic cases with positive PST who underwent the DIPI technique [5] during the second semester of 1988 and the first semester of 1989, with a pregnancy rate of 20.8% per patient. Disappointing results were obtained in the same period with DIPI performed in a dyspermic group with negative PST, with a pregnancy rate of 5.7% per patient.

In the last 3 years our group has obtained better and better results in IVF/ET, with a normospermic pregnancy rate which has risen from 14.3% in 1986 and the first trimester of 1987 to 20.7% per patient in the fourth trimester of 1989. This fact has led us to reconsider our attitude towards the use of semen from dyspermic patients with negative PST in this type of assisted conception. We have been encouraged in this by the possibility of using new sperm treatment techniques such as centrifugation on a discontinuous Percoll gradient [6], both in its classical form and in the variation known as Mini-Percoll.

We decided to use the IVF/ET cases for this study since this is a technique which requires a thorough investigation of the real fertilizing potential of the sperm. We have therefore included in our IVF/ET program of the second bimester of 1990 a group of patients who had previously been excluded from the program for severe dyspermia and negative PST. This group consisted of 101 cases of whom 38 underwent semen treatment on a

Table 3. IVF/ET results (March–April 1990)

	Patients	Type 3,4 sperm	Fertilization rate, %	Preg-nancies	Pregnancy rate, %
NS Pellet swim-up	31	33.7	72.9	6[a]	19.3
DS (PST–) Pellet swim-up	63	2.5	23.6	1[b]	1.6
DS (PST–)[1] Percoll	38	3.1	39.3	5[c]	13.1

[1] In 8 out of 38 patients the Mini-Percoll technique was used: 1 pregnancy in this group.
a vs. b: $\chi^2 = 7.03$ (p $<$ 0.01); a vs. c: $\chi^2 = 3.85$ (p $<$ 0.05).

discontinuous Percoll gradient and 63 with pellet swim-up (PS); the control group was made up of 31 normospermic patients where the semen was treated by PS. Furthermore, in the same period, we performed TEST [7] on 5 normospermic and on two dyspermic cases. In all these cases, the sperm was always treated with PS. The normospermic control group confirmed the pregnancy rates obtained in similar sperm conditions by our group with IVF/ET in the last trimester of 1989. Table 3 shows the pregnancy rate of 19.3% per patient.

We obtained extremely disappointing IVF/ET results in the dyspermic patients with negative PST where the semen was treated with PS, with only one pregnancy (pregnancy rate 1.6% per patient) (table 3). Although oocyte recovery rate was high at 5.5 $\pm$ 3.2 per patient, fertilization rate was extremely low at 23.6%.

The improvement in IVF/ET sperm quality in our group, as shown by the control cases, did not lead to satisfactory results in patients with severe dyspermia when PS was used. Unexpected results were obtained however in cases with severe dyspermia where the semen was treated with centrifugation on a discontinuous Percoll gradient (table 3). We obtained 5 pregnancies, with a rate of 13.1% per patient. It must be borne in mind that in 8 of 38 patients the Mini-Percoll technique was used: one pregnancy was obtained in this group.

Table 3 shows the significant difference between the pregnancy rates ($\chi^2 = 7.03$; p $<$ 0.01) obtained in the normospermic and the dyspermic groups, respectively, with PS sperm preparation. There was, however, no significant difference between the normospermic and dyspermic groups where the semen was prepared by means of centrifugation on the discontinuous Percoll gradient; furthermore, a significant difference was found ($\chi^2 = 3.85$, $p < 0.05$) between the dyspermic cases where the semen was prepared by the PST and those where it was prepared by means of centrifugation on the discontinuous Percoll gradient.

Although our main interest in March and April 1990 was in IVF/ET, at the same time we also performed 5 TESTs (table 4), in couples with unexplained infertility who had previously undergone several cycles of IVC/INSEM [8] and/or DIPI. The pregnancy rate of 4 of 5 (80%) was remarkable, but the number of cases treated was too low for valid conclusions to be drawn and would probably be reduced as the number of cases increased. What is more important is that it seems extremely clear that intrauterine embryo transfer has a negative effect on pregnancy rate in IVF programs. We are thus of the opinion that TEST is the most efficient method of assisted conception in cases of male subfertility. TEST compared to other techniques such as GIFT and PROST has the enormous advantage of involving a thorough study of the real fertilizing potential of the sperm, and at the same time does not seem to be particularly affected by the limits of intrauterine embryo transfer. Only 2 of the patients undergoing TEST in this series belonged to dyspermic couples – and what is more with positive PST and therefore not very severe cases; it is, however, extremely encouraging that in both these cases a pregnancy was obtained (table 4).

The basic question to face is why Percoll-treated sperm manages to increase pregnancy rate so significantly in dyspermic couples undergoing IVF/ET compared to semen treated with the PST technique.

The main drawback of PS is that it does not lead to a truly qualitative sperm selection. The sperm with good motility are generally, but not always, those with high fertilizing potential, for example, the so-called 'pinhead forms'. PS makes it possible to obtain a sufficient number of sperm, but these may not necessarily be of good quality, whereas the Percoll technique produces both sperm quantity and quality. Table 5 shows how it is performed. Without going into a detailed explanation of the method, we should like only to emphasize that the sperm found at the bottom of the test tube are not only more motile, but also have better morphology, since during centrifugation they arrange themselves parallel to the direction of the centrifugal force with their heads, which are their

Table 4. Test results (March–April 1990)

Cases	Type 3,4 sperm	Harvested oocytes	Fertilized oocytes	Transferred embryos	Pregnancy
1	40	2	1	1	not
2	60	7	5	5	yes
3	35	7	4	4	yes
4	40	9	9	6	yes
5	25	2	2	2	yes
6	6	6	2	2	yes
7	10	8	2	2	yes

Table 5. Sperm separation by Percoll gradient

The 100% Percoll solution is diluted with ZA 148 into solution of 90, 80, 70, 55 and 40%; the gradient is formed by placing each Percoll solution in a tube

Sperm is placed on the upper part of the gradient (40%); centrifugation at 250 rpm for 15 min; aspiration to eliminate the 40–90% fractions; washing with 5 ml of medium followed by centrifugation at 250 rpm for 5 min of the 90% fraction; elimination of supernatant after which the pellet is resuspended

heaviest part, towards the vector; the sperm will therefore move towards the centrifugal force according to their own progressive speed and after the necessary period of centrifugation they can be collected from the various densities of the discontinuous gradient depending on the distribution of their relative velocity.

A careful observation of the results of the 5 successes with IVF/ET in dyspermic patients treated with the Percoll technique is a clear indication of the qualitative aspect of this method of semen selection. Table 6 shows that only two of these 5 patients showed type 3 (forward progressive) and 4 (rapidly forward progressive) motility in the untreated ejaculate.

Another case showed only sperm with type 1 (in loco) and type 2 (sluggish, forward, nonprogressive) motility; in the other 2 patients there were only types 2 and 3 sperm. The Percoll treatment in these three cases probably selected a large quantity of good quality sperm from the few sperm present which possessed both adequate motility *and* morphology, and were therefore capable of fertilizing the oocytes.

Table 6. Utilization of discontinuous Percoll gradients in dyspermic patients treated by IVF-ET in 5 cases in which pregnancy was obtained

Patient No.	Motile sperm concentration in nontreated ejaculate after 60 min	Harvested oocytes	Fertilized oocytes	Transferred embryos
1	2 (3.4)	4	3	3
2	10 (3.4)	13	4	4
3[1]	10 (1.2)	8	5	5
4	26 (2.3)	4	3	3
5	6 (2.3)	9	5	5

[1] Mini-Percoll.

Table 7. IVF/ET: fertilization rate and transfer rate/patient (March–April 1990)

	Patients	Fertilization rate, %	Transfers	Transfer/ patient, %
NS Pellet swim-up	31	72.9	30	96.7
DS (PST–) Pellet swim-up	63	23.6	32	50.7
DS (PST–) Percoll	38	39.7	31	81.5

Another indirect proof of the qualitative nature of the selection with the Percoll method is that an oocyte fertilization rate which was not particularly high at 39.7% subsequently led to a fairly high transfer rate per patient (81.5%) (table 7). Every so often this type of treatment has made it possible to select an aliquot of sperm with adequate fertilizing potential from a quantity which was not high enough to permit a high oocyte fertilization rate, but which *was* enough to guarantee constant, uniform fertilization of a certain number of oocytes in the majority of the patients treated.

In conclusion, it is our opinion that the use of centrifugation on the discontinuous Percoll gradient or the even more modern method with

Nycodenz [9], possibly associated with TEST, will open up new horizons in the treatment of male subfertility by means of assisted conception techniques.

References

1   Cimino C, Cefalù E, Ciriminna R, Barba G, Falcone G, Gabrieli M, Venezia R: Il Pellet Swim-up test nella valutazione del seme per la FIV/ET; in Cittadini E, Quartararo P (eds): Atti del XIV Congresso Nazionale della Società Italiana di Fertilità e Sterilità, 1988, pp 281–285.

2   Salerno P, Cimino C, Barba G: Valutazione e trattamento del seme nelle tecniche di fecondazione assistita: Esperienza del centro FIV/ET-GIFT di Palermo; in Cefalù E, Ferraretti AP, Gianaroli L, Palermo R, (eds): Evoluzione delle tecniche FIVET e GIFT. XVI Corso SIFES-Sirmione, 13–14 Maggio 1988, Incontri Serono, pp 121–136.

3   Cittadini E, Cimino C: Dyspermia and GIFT: Results; in Capitanio GL, Asch RH, De Cecco L, Croce S (eds): GIFT: From Basics to Clinics. Serono Symp Publ from Raven Press, 1989, vol 63, pp 391–397.

4   Cittadini E, Cimino C: Risultati delle fecondazioni assistite nel trattamento dei pazienti con sub-fertilità maschile. X. Corso di aggiornamento sulla fertilità e sterilità S. Margherita Ligure, 4–6 Maggio 1989, CO.FE.SE. Eds, in press.

5   Cimino C, Guastella G, Comparetto G, Gullo D, Perino A, Benigno MA, Barba G, Cittadini E: Direct intraperitoneal insemination (DIPI) for the treatment of refractory infertility unrelated to female organic pelvic disease. Acta Eur Fertil 1988;19: 61.

6   Salerno P, Cimino C, Barba G: Gradienti di Percoll e Swim-up da Pellet: 2 tecniche di perparazione del liquido seminale a confronto; in Cittadini E, Quartararo P (eds): Atti del XIV Congr Nazionale della Società Italiana di Fertilità e Sterilità, 1988, pp 293–296.

7   Yovich JL, Rohini Edirisinghe W, Yovich JM, Cummins JM, Spittle JW: Dyspermiae and GIFT or PROST and TEST; in Capitanio GL, Asch RH, De Cecco L, Croce S (eds): GIFT: from Basics to Clinics. Serono Symp Publ from Raven Press, 1989, pp 411–424.

8   Perino A, Cimino C, Catinella E, Barba G, Cittadini E: In vitro sperm capacitation and intrauterine insemination (IVC-INSEM): A simple technique for the treatment of refractory infertility unrelated to female organic pelvic disease. Clinical results and immunological effects: A preliminary report. Acta Eur Fertil 1986;17:325.

9   Gellert-Mortimer ST, Clarke GN, Gordon Baker HW, Hyne RV, Johnston WIH: Evaluation of Niycodenz and Percoll density gradients for the selection of motile human spermatozoa. Fertil Steril 1988;49:335.

Dr. C. Cimino, Istituto Materno-Infantile, Università di Palermo,
Via Cardinale Rampolla 1, I–90100 Palermo (Italy)

Colpi GM, Pozza D (eds): Diagnosing Male Infertility.
Prog Reprod Biol Med. Basel, Karger, 1992, vol 15, pp 210–221

Technical Strategy for Preparing IVF Sperms: A Computer-Assisted Comparative Study and Results of the Use of Subnormal Semen for IVF

P. Clédon, B. Müller, M. Hamori, H.-R. Tinneberg

University Women's Hospital, IVF-Laboratory, University of Tübingen, FRG

Many studies and investigations have been done to prepare sperms for IVF with an optimal efficiency [1–8]. Throughout our experience in the IVF clinic we could attest that the earliest method of obtaining motile spermatozoa, i.e. the swim-up technique, leads to a poor outcome in the case of subnormal semen. This simple technique is good for normal semen samples, but for oligoasthenoteratozoospermia (OAT syndrome) of different grades it is extremely limited by the very low recovery rates of motile sperms.

Considering our increasing rate of 'male factor infertility' nearly reaching 50% in our IVF program, we have developed two other in vitro approaches to improve the sperm quality suiting for IVF or AIH; the Percoll discontinuous density gradient and the glass bead column (GBC). This article gives the results of a comparative study of the 3 precited methods, performed with a fully automated computer-assisted sperm analyzer: the HTM 2030 from Hamilton Thorn Research. The purpose of this study is to propose a strategy for preparing the sperms according to the quality of the semen and define the limits of each method in regard to different measured parameters. The results of this strategy obtained during 1 year in our IVF program will be analyzed and discussed.

Material and Methods

Subject Selection and Semen Acquisition

Semen samples from 60 patients were used in this study. The samples were obtained by masturbation at the IVF clinic afte 3–4 days of abstinence. Patients were selected randomly and classified according to specific semen criteria. This classification does not take in account the morphology of the spermatozoa. This approximative classification of normal and subnormal semen is performed with regard to the criteria shown in table 1.

Thus, we have selected 3 groups: group 1 for the normozoospermia (WHO criteria); group 2 for the oligoasthenozoospermia, and group 3 for the severe oligoasthenozoospermia with most of the time a high rate of teratozoospermia.

Sample Processing

Semen samples were allowed to liquify at 37 °C and were analyzed within 60 min.

Each ejaculate after liquefaction was submitted to a subjective estimation of the 3 above precited parameters (table 1) for classification and immediately compared with the objective analysis of the HTM 2030.

These values are given in the results as 'native' (see Results native). Then the sample has been washed one time with Ham's F 10 (Seromed F 0715) culture medium supplemented with BSA 8 mg/ml. After centrifugation (5 min at $500\,g$) the pellet was resuspended in the same medium. This suspension was equally shared in 3 parts to be submitted to the 3 preparation procedures. Thus, the 3 procedures were applied on a same semen sample, which allowed an absolute comparative analysis.

Previously, an aliquot was taken to give an absolute 'control' (see Results 'control'); this aliquot has never been submitted to any preparation. After each preparation procedure, the resulting sperm suspension was submitted 3 times to the HTM 2030 for extensive analysis. First, immediately at the end of the procedure, this measure is called 'initial' (see Results 'initial'), then 3 h later (see Results 3 hours) and finally 24 h after the procedure (see results '24 h').

Table 1. University Women's Hospital Tübingen IVF Lab., male factor infertility: guidelines for the classification of patients

	Group 3 oligoasthenospermia (severe) grade III	Group 2 oligoasthenospermia grade I and II	Group 1 normal sperm sample WHO criteria
Sperm density, millions/ml	5	10	20
Percentage motile sperms	10	25	40–50
Percentage progressive sperms	5	15	30

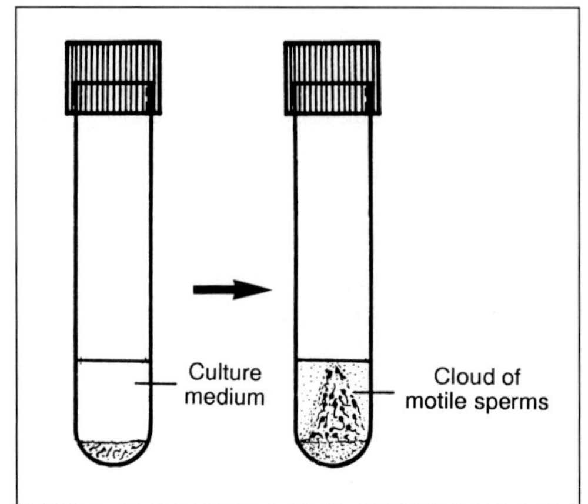

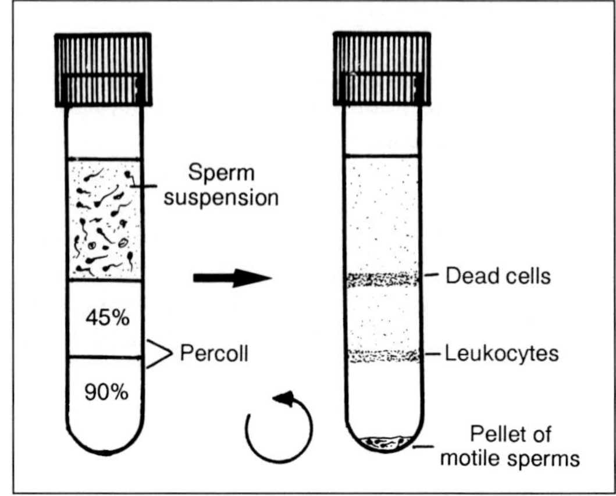

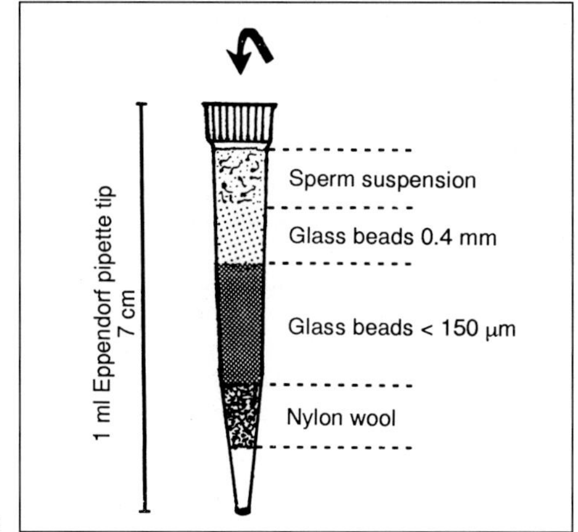

Fig. 1. Sperm preparation procedures:
a Swim-up technique. 30–60 min at 37 °C, 5% CO_2.
b Percoll discontinuous density gradient. 20 min at 200–300 g.
c Glass beads column set-up (GBC).

Sperm Preparation Procedures

Swim-Up Technique. The intial sperm suspension (after first washing step) was once more centrifuged (5 min at 200 g) and the resulting pellet overlaid with the culture medium (Ham's F 10 + BSA). Sperms were allowed to migrate for 30–45 min in groups 1 and 2 and 60 min in group 3. The supernatant was carefully removed and immediately submitted for analysis. Figure 1a shows the principle of this method.

Percoll Discontinuous Density Gradient. A 90% and 45% bilayer of Percoll (Pharmacia) in Ham's F 10 + BSA (Sigma A 7409) was used. After layering the Percoll, the sperm suspension was loaded on top of the gradient as shown in figure 1b. After a centrifugation of 20 min at 200 g, the pellet at the bottom of the tube was removed, washed and submitted for analysis.

Glass Bead Column Filtration (GBC). We used a modified column previously investigated by Daya et al. [3]. Figure 1c shows the principle of the filtration and the composition of the column. The finest glass beads (< 150 μm) are from Sigma 62381 and the upper bigger one from BDH Lab. No. 150.29. The nylon wool comes from Fenwal Lab. USA, code 4C2906.

Semen Motility Analyzer: The HTM 2030

The HTM uses a combination of components to automatically determine sperm motility and concentration. An infrared beam is used to acquire images of the sample. The integrated optical system of the HTM produces a dark field image of the specimen, thus sperm cells appear as bright objects on a dark background. An electronic image is produced and transmitted to the processing system for analysis. The sample is analyzed by collecting successive images at equally spaced time intervals. The HTM has an automated and heated stage for the Makler chamber. For all sperm samples 13 groups of measures were performed. Each group of measures includes the determination of many parameters. We selected 7 parameters (among 17) which appeared to us as very helpful and significative for this comparative study: concentration of total cells, concentration of motile cells, total motility, progressive motility, pathologic velocity, progressive velocity and velocity distribution. Most of these parameters could not be objectively and exactly determined by a well-trained operator.

Statistical Analysis

The data obtained in this study were computer analyzed to allow comparisons between groups. Since sperm densities, motilities and velocities follow a skewed distribution, comparisons were carried out with the use of the Wilcoxon rank-sum test (multiple testing, Holm method for error correcting).

Results

Group 1 (Normozoospermia)

Our results shown in figure 2a–d clearly support the fact that the swim-up procedure represents the most effective method to treat this semen quality. This technique brings an acceptable but lower recovery rate of motile cells (around 20%) than with the Percoll gradient (around 54% higher). In terms of concentration, the GBC is still more efficient (fig. 2a).

As for the quality, i.e. motility (total and progressive), the swim-up gives the best results with, respectively, 79 and 70% and velocities of 63–55 μm/s (path-progressive velocity) (fig. 2b, c). The average total motility and progressive motility in a normal semen sample was significantly higher after treatment with the swim-up technique ($p < 0.05$ for the total motility, $p < 0.01$ for the progressive motility) than after Percoll gradient procedure. At the end of the 24-hour incubation period, the total motility after swim-up is reduced as compared to the Percoll whereas the pathologic and

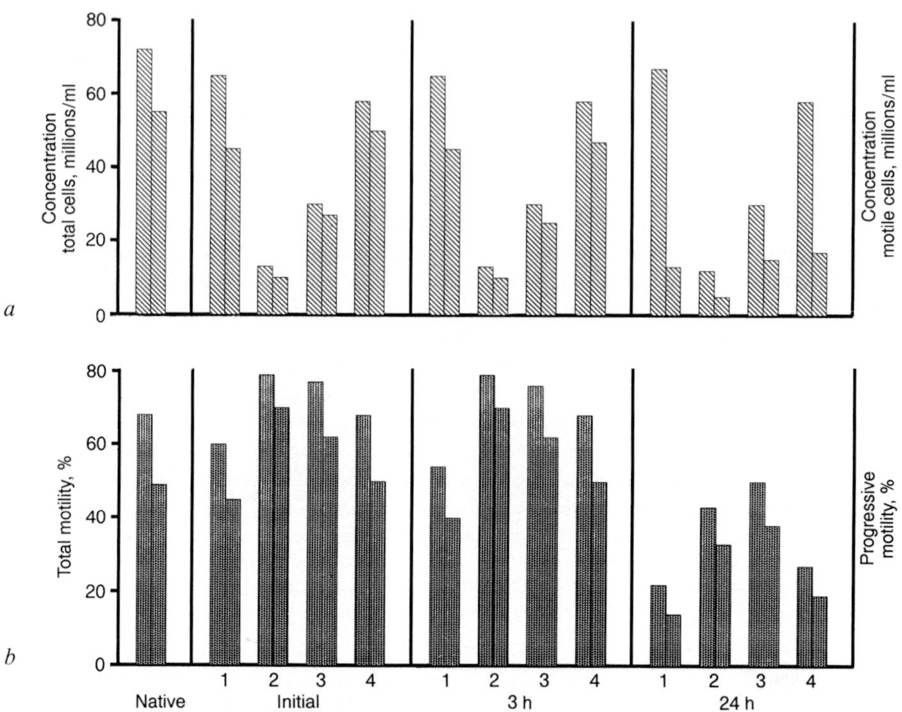

Fig. 2a–d. Results obtained for the normozoospermic group. *a* Concentration of total and motile cells in millions/ml; *b* total and progressive motility in %; *c* path and progressive velocity in μm/s; *d* distribution of the velocities in %. ☐ = rapid (V > 75 μm/s); ▦ = moderate (10 < V < 75 μm/s); ▧ = slow (V < 10 μm/s); ■ = static. 1 = Control; 2 = swim-up; 3 = Percoll gradient; 4 = GBC.

progressive velocity show no significant differences. Figure 2d gives the distribution of velocity: 30% of rapid sperms, 54% of moderate, 4.5% of slow and 13% of static after swim-up. Without detailing the results, the GBC gives for this group the worst results.

For all the 3 procedures these results are confirmed and do not change after 3h of culture, but 24h after the end of the preparation a slight decrease of all parameters is noticed.

Group 2 (Oligoasthenozoospermia – Slight to Moderate)

In this group, the Percoll density gradient brings significantly better results than those obtained after swim-up (fig. 3a–d). Percoll procedure

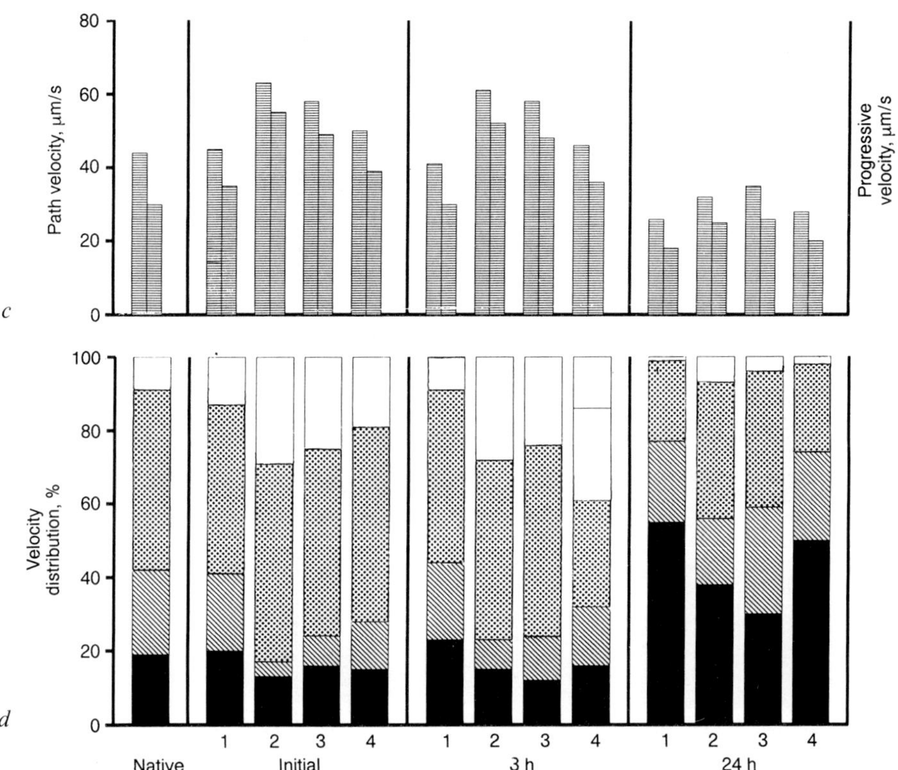

brings significantly higher sperm recovery than after swim-up: 12.4 and 5.6 M/ml instead of 2.7 and 2 M/ml with swim-up for concentration of total cells ($p < 0.001$) and the concentration of motile cells respectively ($p < 0.001$). The motility parameters are comparable but not significantly different, 55 and 43% instead of 61 and 53.6% after swim-up for the total and progressive motility.

Same similarity with the path and progressive velocity: 44 and 36 µm/s for Percoll, and 48 and 41 µm/s after swim-up. These results are not significantly different.

The velocity distribution (rapid, moderate, slow, static) gives, respectively, 9, 46, 12 and 33% after Percoll and 10.8, 50.6, 6.5 and 32.4% after swim-up. After 3 h in culture no significant difference in all the parameters is to be noticed. At 24 h some differences are appearing in favor of the Percoll procedure. Percoll-treated sperms keep a higher motility (total/

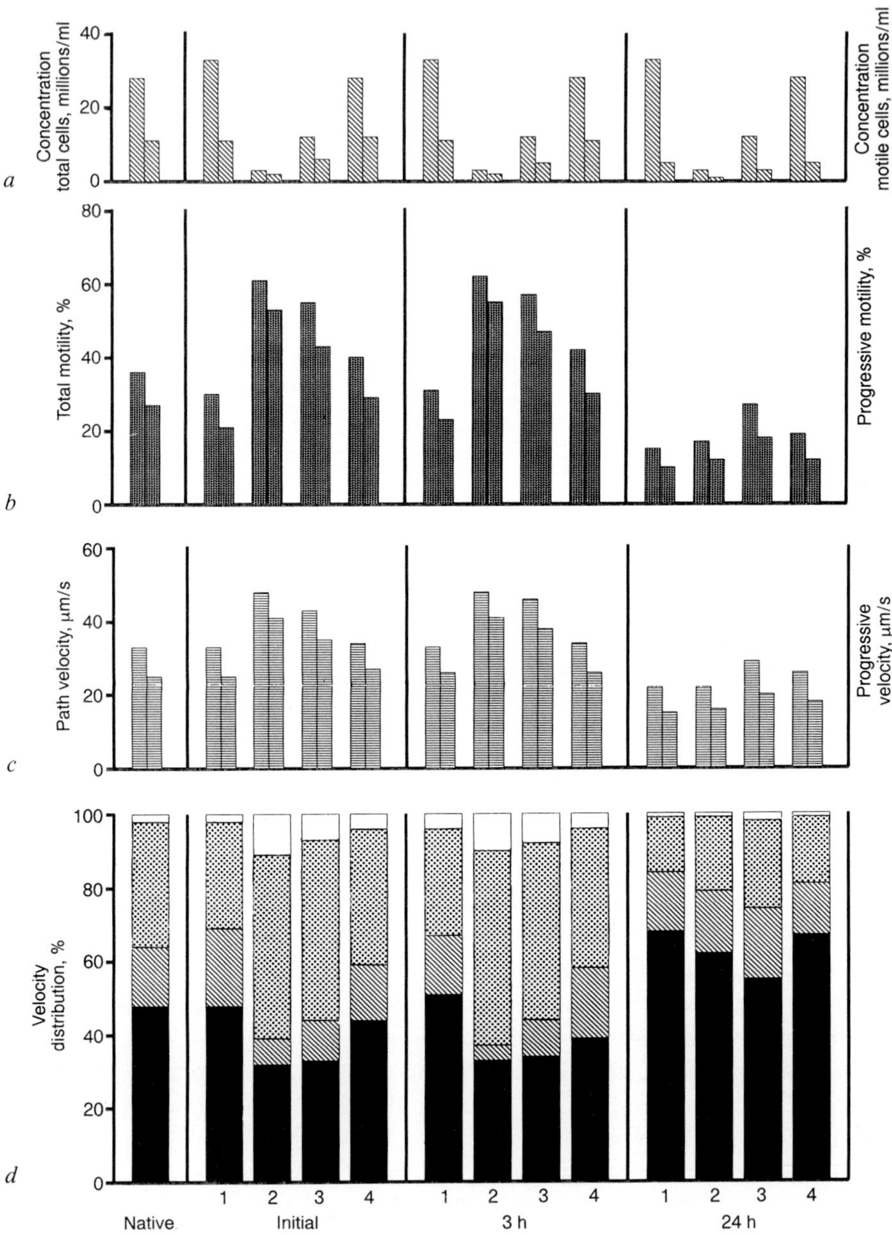

Fig. 3a–d. Results obtained for the oligoasthenozoospermic group. For *a-d*, figures and symbols see figure 2.

progressive) of 27 and 18%, than after swim-up: 18 and 12%. The same tendency is observable at 24 h for the velocity parameters (path, progressive) 30 and 20 μm/s after Percoll and only 21 and 15 μm/s after swim-up. In this group 2, the GBC brings still higher concentrations of sperms but of a lower quality regarding motility and velocity.

If we examine the distribution of velocity for this group, the proportion of rapid sperms is slightly higher (10.8%) after swim-up than after Percoll gradient (8.9%). But after 24 h the proportions of rapid and moderate sperms remain higher after Percoll (1.4 and 25%, respectively) than after swim-up (0.5 and 17%). These last comparisons were not found to be highly significant. The use of Percoll gradient for this type of pathological sperms gave satisfying results in our IVF program.

With a fertilization rate reaching 60% (compared to 75% after swim-up procedure) we could obtain a 22% pregnancy rate per embryo transfer compared to 19% after swim-up (table 2).

Group 3 (Severe Oligoasthenozoospermia)

In this group the initial concentration of total cells is very low (around 6 M/ml) and still lower for the concentration of motile cells (around 1 M/ml). In terms of quantity once more the GBC gives here significantly ($p <$ 0.05) the best results with 3 M/ml of total cells compared with 1.8 and 0.9 M/ml for Percoll gradient and swim-up (fig. 4a). For all of the other parameters, the GBC provides slightly better results than the 2 other methods.

After 24 h, GBC selected sperms show a good velocity distribution (fig. 4d). After swim-up or Percoll, absolutely no rapid sperms remain in

Table 2. IVF results: male factor infertility (from 1.5.1989 till 30.4.1990)

	Patients	Follicle punction (FP)	Fertilization rate %	ET (per FP %)	Pregnancy (per ET %)	Abortion	Ectopic pregnancy
Swim-up (Group 1)		201	75	137 (68)	26 (19)	1	2
Percoll density gradient (Group 2)		57	60	23 (40)	5 (22)	1	0
GBC (Group 3)		22	36	7 (32)	0 (0)	0	0
Total	215	280		167 (60)	31 (19)	2	2

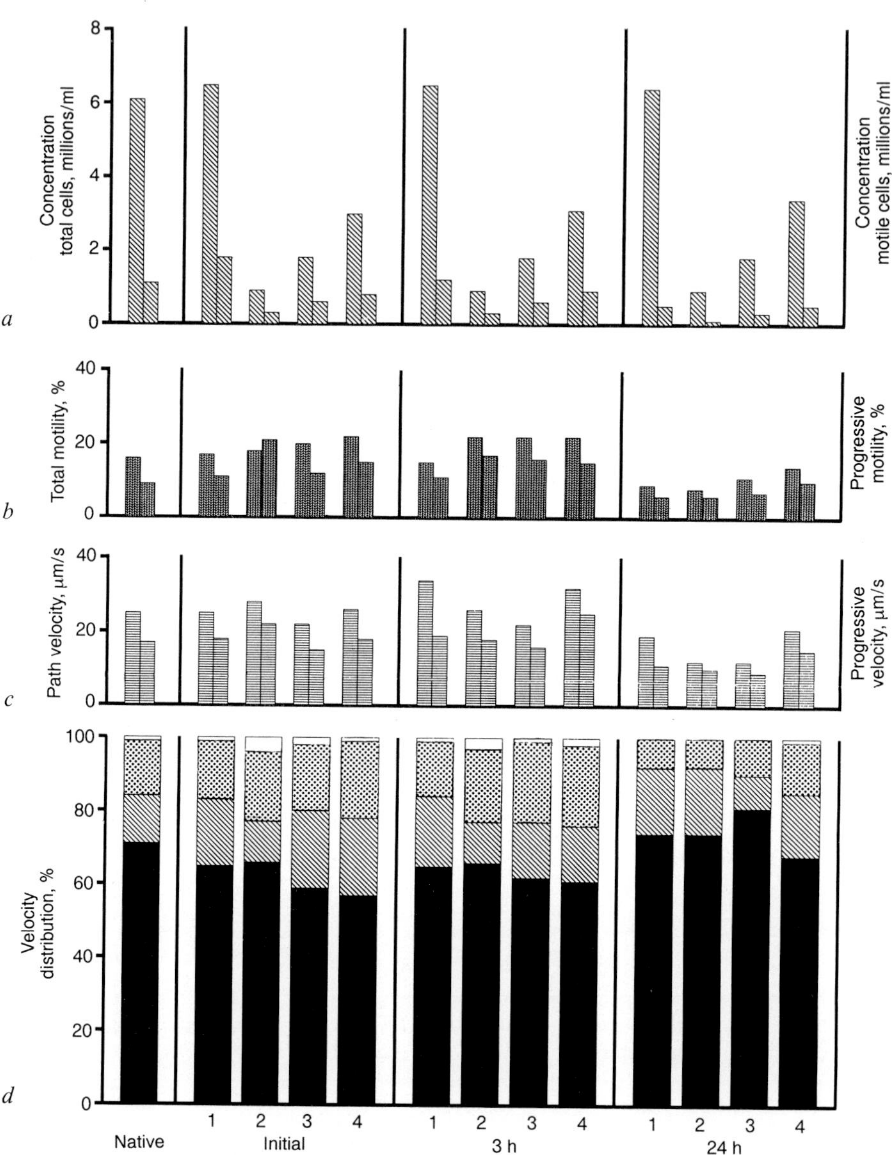

Fig. 4a–d. Results obtained for the severe oligoasthenozoospermic group. For *a-d*, figures and symbols see figure 2.

the culture medium, whereas 0.15 M/ml of rapid sperms still exist after GBC preparation. The proportion of moderate sperms (velocity between 10 and 75 µm, known as good sperms with fertilization potentialities) is higher after GBC than after the other methods (not significant).

Results obtained after 22 treated cycles are shown in table 2. The fertilization rate is low but still acceptable (36%), leading to a non-negligible embryo transfer rate per treated cycle (32%). Unfortunately, no pregnancy has been observed. Possible reasons for this failure will be discussed.

Discussion

It clearly appears that all the 3 methods for preparing the sperms, applied in this study, bring different results in regard to the type of semen and type of observed parameters. These results can be appreciated in a quantitative approach (concentration of total and motile cells) or in qualitative ones (percentage of motile and progressive motile cells, velocities). The recovery of ideal sperms for IVF represents a compromise between these quantitative and qualitative parameters of the corresponding semen profile.

Each method selects the sperms on different criteria and brings different recovery rates for good sperms. The use of the HTM 2030 helped us to objectively and exactly estimate the positive and negative aspects of each method. After examining our results, we can now attest that using only one sperm preparation method in an IVF/AIH unit (making no selection about andrological subfertilities) makes no sense. Our study clearly shows that these 3 easy-to-perform methods are fairly complementary for providing a better outcome of the IVF treatment or at least better hope of reaching some fertilization and consequently embryo transfer.

Each procedure has its own limits as regards semen quality; so we are proposing the following strategy with nevertheless some precautions to be taken.

For our first group (normozoospermia) it is for us quite clear that the swim-up procedure represents the best selection procedure for good sperms. Although the recovery rate of motile sperms (concentration of motile sperm) are always lower than with the other methods, there are enough good sperms to perform an insemination. These good sperms exhibit a better motility (total and progressive), velocity (pathologic, progressive) than those obtained with other procedures.

For group 2, slight-to-moderate oligoasthenozoospermia, the swim-up procedure can also give acceptable results but never sufficient from the quantitative point of view. Sperm suspensions recovered after the swim-up procedure do not exhibit all the satisfying parameters to perform the insemination of many eggs. It appears for us that the Percoll procedure gives the best sperm suspension profile. Moreover, our data suggest that Percoll-treated sperms (group 1 or 2) reveal a higher percentage of motility and progressive motility at 24 h of culture than those treated with the swim-up procedure or the GBC. This means that this Percoll gradient, selecting sperms on their density, provides sperm of moderate quality (moderate motion characteristic) which longer survive in culture.

For group 3, i.e. with severe oligoasthenozoospermia, the GBC is for us the only method which still brings a reasonable concentration of motile sperms. In comparison and considering our experience in the IVF program, sperm suspensions recovered after swim-up or Percoll gradient of this type of semen profile never allow the minimal prerequisite to perform an insemination of even one oocyte.

At 24 h of culture, sperms exhibit a good motility and velocity not significantly different from the control group and higher than sperms prepared with Percoll or swim-up. It is important to note that in this group the percentage of teratozoospermia is always very high (greater than 70%). The GBC not being a highly selective procedure for the sperms according to their progressive motility, means that nearly only living and moving sperms are running through the column, including the abnormal forms. In this way, our observations do not exactly confirm the results obtained by Daya et al. [3] using another variant of column. Moreover, the final cell suspensions are never totally free of leukocytes.

These findings explain the poor fertilization rates (36%) (table 2). Therefore, we advise to use this procedure only when the percentage of subnormal sperms does not exceed the 60–70% limit. We have applied this technique in our IVF program with some consistent modifications of the traditional way to culture and inseminate the oocytes. Shortly after the follicle punction, the follicles are briefly treated with hyaluronidase to remove the cumulus cells. The corona radiata is left partially intact. The oocytes free of their cumulus cells are then set into microdrops of culture medium (10 µl) under paraffin oil.

Microvolumes of sperms are then introduced into the drop. These technical modifications increase the chances of achieving fertilization. This technique brings 32% of embryo transfer rate per follicle punction without being a traumatic method like microinjection of sperms into the perivetellin space.

We could further improve all of these 3 methods by a previous incubation (minimum 1 h) of the washed sperms into Ham's F 10 and BSA containing 10 mM of taurine, 6 mM caffeine or 5 mM of 2-deoxyadenosine. Preincubaton with these chemicals improved the progressive motility by 36%, the percentage of rapid sperms by 20% and the beat frequency by nearly 20% (data not shown). This part of the work still remains under current investigation.

Acknowledgement

We want to thank Mrs. S. Ruhland for her help in the statistical treatment of the data. our study has been kindly supported by the Labotec Company (Göttingen, FRG) and the Nikon Co. (M.M. Joos, Calw, FRG). Thanks are extended to Mrs. K. Luzecki for her technical assistance and Mrs. R. Pompe for typing the text.

References

1   Yates CA, De Kretser DM: Male factor infertility and in vitro fertilization. J In Vitro Fertil Embryo Transfer 1987;4:141–147.
2   Lessley BA, Garner L: Isolation of motile spermatozoa by density gradient centrifugation in Percoll. Gamete Res 1983;7:49–61.
3   Daya S, et al: Separation of motile human spermatozoa by means of a glass bead column. Gamete Res 1987;17:375–580.
4   Berger T, et al: Comparison of techniques for selection of motile spermatozoa. Fertil Steril 1985;43:2.
5   Wikfand M, et al: A self migration method for preparation of sperm for in vitro fertilization. Hum Reprod 1987;2:191–195.
6   Lucerna E, et al: Recovery of motile sperm using the migration-sedimentation technique in an in vitro fertilization – embryo transfer program. Hum Reprod 1989;4:163–168.
7   Andolz P, et al: Improvement of sperm quality in abnormal semen samples using a modified swim up procedure. Hum Reprod 1987;2:99–101.
8   Puntjabi U, et al: Comparison between different pretreatment techniques for sperm recovery prior to intrauterine insemination, GIFT of IVF. Hum Reprod 1990;5:75–83.

Dr. P. Clédon, University Women's Hospital, IVF-Laboratory,
University of Tübingen, Schleichstrasse 4, D-W–7400 Tübingen (FRG)

Colpi GM, Pozza D (eds): Diagnosing Male Infertility.
Prog Reprod Biol Med. Basel, Karger, 1992, vol 15, pp 222–225

Simple Seminal Analysis Is a Useful Predictor for the Outcome of Assisted Conception

An Update

John P. Pryor

St. Peter's Hospital, London, UK

Attempts have been made to standardize the technique of seminal analysis [1] but despite this the results often fail to correlate with the fertility status of the patient. In order to improve the predictive value of seminal analysis, there has been a trend towards more sophisticated computer analysis. Another approach was to test the fertilizing capacity of sperm using hamster oocytes. This paper correlates the result of simple seminal analysis with the pregnancy rates at the time of assisted conception.

Methods

During the period January 1988 to December 1989 inclusive, 750 women underwent 548 cycles of egg collection for gamete intrafallopian tube transfer (GIFT) or 350 cycles of egg collection for in vitro fertilization (IVF) at the Lister Hospital in London. The techniques utilized have been described elsewhere [2] and seminal analysis was performed according to WHO guidelines although motility was assessed differently (table 1). Pregnancy was confirmed by chemical testing and ultrasound examination. Women with tubal obstruction were unsuitable for GIFT but otherwise there was no difference in the distribution between GIFT and IVF in the different categories of patients undergoing assisted conception. The pregnancy rates for IVF and GIFT have therefore been aggregated for the subsequent analysis.

Results

The overall pregnancy rates were 31 % for GIFT and 23 % for IVF. The category of infertility influenced the pregnancy rate and the lowest rates were in those patients where the semen quality was subnormal according to

Table 1. Assessment of the quality of progressive sperm motility

WHO 1980		WHO 1987		Lister	
		d	immotile	0	
0	none	c	none	1	none
1	poor (weak)	b	slow or sluggish linear or nonlinear	2	slow/sluggish
2	good (moderate)	a	rapid and linear	3	forward
3	excellent (very active forward)			4	excellent forward

Table 2. Pregnancy rates for IVF and GIFT according to the different causes for infertility

Cause	Number	Pregnant
Unexplained	381	37%
Endometriosis	30	37%
Tubal obstruction	250	24%
Anovulation	27	22%
Male factor	195	17%

the WHO standards (table 2). There was no difference in the pregnancy rates when normal husband's semen was used (31% in 603 cycles) or with the use of donor semen (33% in 62 cycles).

Analysis of the various male factors in relation to pregnancy rates showed good pregnancy rates in relation to those men with antisperm antibodies (43% in 7 cycles), moderate rates when the sole abnormality was teratozoospermia (25% in 20 cycles) and no pregnancy in the seven attempts at sperm retrieval for assisted conception in obstructive azoospermia as originally described [3]. The results in these patients are not analysed further and the subsequent analysis will be related to the variables of sperm number, sperm motility and the quality of sperm motility.

When the semen abnormality was solely reduction of sperm numbers below 20 million/ml (45 men) then the pregnancy rate was 18%, when the sperm motility was below normal (56 men) the pregnancy rate was 12.5% and for men with reduced sperm numbers and motility (54 men) the preg-

Table 3. Pregnancy rate for IVF/GIFT related to the motile sperm concentration

Motile sperm concentration million motile/ml	Number	Pregnant
<1	22	5%
1–4	42	12%
5–9	34	21%
10–14	21	14%
15–19	11	18%
>20	30	20%
Normal	665	31%

Table 4. Relationship between the quality of progressive motility and pregnancy rate

Quality of movement	Sperm concentration			
	4 m/ml or less		more than 4 m/ml	
	number	pregnant, %	number	pregnant, %
1	4	0	4	0
1–2	7	0	3	0
2	21	10	13	15
2–3	26	12	50	22
3	11	9	24	20

nancy rate was 19%. Table 3 gives the pregnancy rates with regard to the number of motile sperm (million/motile/ml) and shows that it was only when there was less than 5 million motile sperm/ml that the pregnancy rate fell within the group of men with abnormal sperm. It is also of interest to relate the quality of sperm motility to the chances of pregnancy (table 4).

Discussion

Seminal analysis is performed in an attempt to define the fertility status of the male partner. It is important to know the predictive value of the results obtained and whilst azoospermia is certainly associated with

Table 5. Pregnancy rate for women of different ages

Age, years	Number	Pregnant, %
20–29	175	41
30–34	311	29
35–39	255	25
40 and over	124	15

impaired fertility, other values are less meaningful. It is important to recognize that it is the number of 'good' sperm in the ejaculate that is critical rather than the total number of sperm regardless of their quality. It is not surprising to find that pregnancy did not occur unless there was some progressive motility (Lister grade 2, WHO grade b). Reduced pregnancy rates (10%) occurred when the total number of motile sperm was less than 4 million/ml regardless of the quality of movement. The sensitivity of the semen test to detect infertility is poor although its specificity at that level was useful as a screening test (97%). Wherever the semen quality was reduced below the WHO guidelines, the pregnancy rate was reduced from 30 to 20%. It should be remembered that even with assisted conception the female factors are of importance. The pregnancy rate declined from 41% in those women aged less than 30 years to only 15% for those over 40 years of age (table 5).

References

1   World Health Organisation: WHO Laboratory Manual for the Examination of Human Semen and Semen-Cervical Mucus Interaction, ed 2. Cambridge, Cambridge University Press, 1987.
2   Abdalla HI, Ahuja KK, Leonard T, Morris NN: The value of IVF-ET in patients undergoing treatment by the GIFT procedure. Human Reprod 1988;3:944–947.
3   Pryor JP, Parsons JH, Goswamy RK, Matson PL, Vaid P, Wilson L, Whitehead MI: In vitro fertilisation for men with obstructive oligozoospermia. Lancet 1984;ii: 762.

Dr. John P. Pryor, St. Peter's Hospital, Henrietta Street,
London WC2E 8NE (UK)

Colpi GM, Pozza D (eds): Diagnosing Male Infertility.
Prog Reprod Biol Med. Basel, Karger, 1992, vol 15, pp 226–231

Male Infertility and in vitro Fertilization

*H.G. Massouras, M.E. Sakellariou, D. Kollia, G. Androutsos,
K.L. Oikonomou*

IVF Centre, 'Agios Panteleimon', Athens, Greece

Although originally intended for the treatment of female factor infertility, in vitro fertilization and embryo transfer (IVF-ET) has been extended to include couples with male factor as well. In the present study we deal with 50 cases (patients/cycles) of infertile couples after 5–10 years of normal sexual life whose infertility was exclusively due to the male factor. Most of them had already tried unsuccessfully other methods of fertilization (AIH, AID, or GIFT) depending on the male seminal analysis.

We have analyzed the effects of the conventional semen parameters of sperm concentration and motility on fertilization rates over the last 15 months. The other semen parameters (ejaculate volume, liquefaction time, morphologic features, etc.) were examined but are not considered in this study.

L-Carnitine, fructose, ions were also routinely examined. An effort has also been made to detect the causative factors of male infertility and the rate of their relation to the presence of the two semen parameters of low concentration (oligospermia) and low motility (asthenospermia). A history of infectious parotitis (mumps), tuberculosis and infections of the urogenital tract were found responsible for a high rate of low sperm count and low motility (75%). Diabetes, diseases of the thyroid gland, congenital genital abnormalities and other systematic diseases formed a second group (5%). To a smaller extent, we had diseases attributed to certain professions (conditions of work in excessive heat, chemical environment mostly responsible for teratogenesis, etc.) and cases of varicocele. Finally, external factors, such as smoking, alcohol, living conditions and other environmental causes, had a (5%) contribution to the male infertility problem.

Table 1. Sperm concentrations in groups A–C

	Motility $< 40\%$	Motility $> 40\%$
Group A: Sperm concentration $<10 \times 10^6$/ml	100%	–
Group B: Sperm concentration between 10×10^6/ml and 25×10^6/ml	3/15 cases: 20%	12/15 cases: 80%
Group C: Sperm concentration $>25 \times 10^6$/ml	–	100%

Materials and Methods

The 50 couples were divided into 3 groups according to the sperm concentration and motility. The first group (group A) consists of patients with sperm concentration $<10 \times 10^6$/ml and motility $< 40\%$. Patients of group B had a concentration between 10 and 25 $\times 10^6$/ml (motility over or below 40%) and group C had a concentration more than 25×10^6/ml and motility $> 40\%$ (table 1). We did not use donor sperm even in cases with very poor semen samples.

There were two protocols for semen preparation for IVF. The first was used for semen samples with a count less than 10×10^6/ml (group A) or in some cases of group B with very poor motility (3 cases), whereas the second protocol for both group B and C.

Semen samples were obtained from the husbands 2–3 h prior to insemination and the semen was analyzed immediately after liquefaction for the conventional semen parameters [1]. When liquefaction failed to occur within 30 min or the semen remained highly viscous, syringing through a 21-gauge needle reduced viscosity and enabled more accurate analysis. Semen was diluted with an equal volume of culture medium + 10% IMS and washed twice or once for very poor sperm, by centrifugation at 200 g for 12 min the first time and 6 min the second time (table 2). After the final wash, the supernatant in the first protocol was decanted and the sperm pellet was resuspended and divided in the 4 wells of a 4-well Multidish (Nunc, Kamstrup, Roskilde, Denmark) (table 3). Fresh culture medium (1 ml) was gently layered over the pellets and the Multidish was placed in the incubator at 37 °C for 1–2 h depending on the sperm concentration and motility. The top half or more of the culture medium containing the highly motile spermatozoa was then transferred to a 5-ml tube (Falcon Plastics, Oxnard, Calif., USA) and centrifuged till the final concentration reached motile sperm/ml.

In the second protocol after the final wash and the decantation of the supernatant, the sperm pellet was resuspended in the remaining culture medium in the tube by gentle shaking. Fresh culture medium (2–3 ml) was gently layered over the resuspended pellet and the tube was placed in the incubator at 37 °C for 20–40 min (table 3).

The top 1 ml of the medium containing the highly motile spermatozoa was transferred to another tube and centrifuged or used immediately for insemination if the concentration was the desirable (about 1×10^6/ml). For the insemination we used 100,000–150,000 motile spermatozoa/ovum when the sperm motility was poor, whereas no more than 100,000 were necessary in samples with good motility. The insemination took place

Table 2. Sperm preparation

1	Liquefaction for 30–40 min at 37 °C
2	Count and analysis of semen parameters
3	Dilution of semen with an equal volume of culture medium + 10% IMS
4	Centrifugation for 12 min at 200 g
5	Decantation of the supernatant
6	Centrifugation of the pellet for 6 min at 200 g
7	Decantation of the supernatant

Table 3. First and second protocols

	First protocol	Second protocol
8	Dilution of the pellet in a 4-well Multidish	–
9	Placement of 0.5–1 ml of new culture medium in every well of the 4-well Multidish	Placement of 2–3 ml of new culture medium on the pellet (in the tube)
10	Incubation for 1–2 h	Incubation for 20–30 min
11	Replacement of the swim-up	Replacement of the swim-up
12	Count and percentage of motility	Count and percentage of motility
13	Usual centrifugation of the swim-up	Rare centrifugation of the swim-up
14	Insemination with 100,000–150,000 motile spermatozoa/ml	Insemination with up to 100,000 motile spermatozoa/ml

4–6 h after the egg recovery for the mature eggs and after 24–28 h for the immature ones [2, 3].

The criterion of fertilization was the presence of two pronuclei 16–18 h after insemination. If fertilization had occurred we placed the embryos to new culture medium with 20% IMS. The embryo transfer was performed about 32–36 h later.

Results

In this study it is clear that when the parameters of semen quality were below the normal fertile range (sperm concentration below 10 × 10^6/ml and motility < 40% or concentration < 25 × 10^6/ml combined with low motility) there was a significant reduction in the fertilization rate. In these

Table 4. Fertilization data

	Group A	Group B	Group C
Number of cases	12	15	23
Fertilization rate/cases	2/12 (16.7%)	10/15 (77%)	21/23 (91.3%)
Number of oocytes fertilized/insemination	10/47 (21%)	28/55 (51%)	49/68 (72%)
Number of embryos with good development/insemination	7/47 (14.8%)	14/55 (25.5%)	27/68 (40%)
Pregnancies	1/12 (8.3%)	3/15 (20%)	8/23 (35%)
	1 ongoing	1 ongoing	2 deliveries
	(4th month)	(5th month)	4 ongoing
		2 aborted	(3rd, 4th, 7th)
			2 aborted

cases the quality of the sperm (mainly motility) proved to be more important than the quantity in the fertilization rate. In group A (table 1), where the concentration was $<10 \times 10^6$/ml, motility was $<40\%$ in all cases, whereas in group B, where the concentration was between 10 and 25×10^6/ml, motility was $<40\%$ in 3/15 cases (20%) and $>40\%$ in 12/15 cases (80%). In group C (concentration $>25 \times 10^6$/ml) motility was $>40\%$ in all cases.

In group A (table 4) fertilization was achieved in only 2 cases (16.7%) followed by a very low percentage of the fertilized oocytes (21%). In the remaining 10 cases no fertilization was possible despite the good quality of oocytes. In group B we had a remarkable increase in the fertilization rate (77%). The number of oocytes fertilized and of those with good development showed a correspondingly satisfactory increase as well. In 2 out of the 5 unfertilized cases of this group, motility was extremely poor (sluggish movement). In the remaining 3 cases lack of fertilization was mainly attributed to the presence of antisperm antibodies. In these cases we proceeded to desensitization according to Mowbray [4, 5]. In group C, fertilization rate was very high (91.3%) followed by an analogous percentage of pregnancies in the number of oocytes fertilized. In the 2 unfertilized cases of this group the presence of antisperm antibodies was again responsible (as it happened in group B). The number of pregnancies achieved in each group followed a fairly proportionate pattern (table 4).

Discussion

The present study shows that sperm concentration and motility are the main and most significant parameters of semen analysis for the success of IVF. Sperm concentration played an important role in increasing fertilization rates, but only when accompanied with satisfactory motility (about or over 40%), as it usually appeared in sperm samples with $>10 \times 10^6$/ml count. Low sperm concentration ($<10 \times 10^6$/ml) was always accompanied with poor sperm motility ($<40\%$).

The application of the first protocol (in group A) enabled us to harvest as many spermatozoa (100,000–150,000 motile spermatozoa/ovum) as needed for insemination, although the sperm count was, in some cases, very poor. But even though we achieved the desirable number of motile spermatozoa/ovum, fertilization occurred only in 2 cases with 4×10^6 and 9×10^6/ml total motile sperm/ejaculate (16.7%).

At this point, I would like to report a remarkable case which entered our IVF program after the completion of this study. In this case, fertilization has been achieved with 500,000 total motile spermatozoa/ejaculate and only 40,000 motile spermatozoa/ovum. So far, a birth has been reported with as few as 600,000 total motile sperm/ejaculate [6], and a pregnancy with 1×10^6 total sperm in the ejaculate [7]. For semen samples $>10 \times 10^6$/ml and motility $>40\%$ we used up to 100,000 motile spermatozoa/ovum and fertilization occurred in 31/38 cases (82%). We did not use a very low count of motile spermatozoa/ovum even though fertilization has been reported with as many as 10,000 motile spermatozoa/ovum thus reducing the risk of polyspermic fertilization [8, 9]. On the other hand, it has also been already reported by Mahadevan and Trounson [10] that there is a significant reduction in the fertilization rate with $>200,000$ motile spermatozoa/ovum. We must emphasize that IVF has a high success rate in infertile cases, with male factor infertility, where other techniques (AIH, AID, GIFT) have already failed, when sperm concentration is $>10 \times 10^6$/ml and motility $>40\%$ (pregnancy rate 55%). But we must also emphasize that patients with low sperm concentration and poor motility should be informed about the likely success of IVF.

References

1   Mahadevan MM, Baker HWG: Assessment and preparation of semen; in Wood C, Trounso AO (eds): Clinical in vitro Fertilization. Berlin, Springer, 1984, pp 83–97.

2 Trounson A: Factors influencing the success of fertilization and embryonic growth in vitro; in Edwards RG, Purdy JM (eds): Human Conception in vitro. London, Academic Press, 1982, pp 201–205.

3 Edwards RG, Fishel B, Purdy JM: In vitro fertilization of human eggs in vitro. Analysis of follicular growth, ovulation and fertilization; in Beir HM, Linder HR (eds): Fertilization of the Human Egg in vitro: Biological Basis and Clinical Applications. Berlin, Springer, 1988, pp 169–188.

4 Mowbray J: Genetic and immunological factors in human recurrent abortion. Am J Reprod Immunol Microbiol 1987;15:138–140.

5 Mowbray J: Immunology of early pregnancy. Reproduction 1988;3:79–82.

6 Mettler L, Baukloh A, Riedel HH: Immunological and other problems with human spermatozoa; in Edwards JG, Purdy GM (eds): Human Conception in vitro. Orlando, Academic Press, 1982, p 189.

7 Hirsch I, Gibbons WE, Lipshultz LI, Rossarik KK, Young RL, Poindexter AN, Dodson MG, Findley WE: In vitro fertilization in couples with male factor infertility. Fertil Steril 1986;45:659.

8 Graft I, McLeod F, Bernard A, Green S, Twigg H: Sperm numbers and in vitro fertilization. Lancet 1981;ii:1165.

9 Byrd W, Wolf DP, Dandekar P, Quigley MM: Sperm concentration dependency in human in vitro fertilization (abstract). Fertil Steril 1984;41:55S.

10 Mahadevan MM, Trounson A: The influence of seminal characteristics on the success rate of human in vitro fertilization. Fertil Steril 1984;42:400.

Dr. H.G. Massouras, IVF Centre, 'Agios Panteleimon',
GR–10676 Athens (Greece)

Colpi GM, Pozza D (eds): Diagnosing Male Infertility.
Prog Reprod Biol Med. Basel, Karger, 1992, vol 15, pp 232–236

The Use of Therapeutic Donor Insemination to Demonstrate the Inadequacy of Present WHO Standards for Normal Motile Density in Distinguishing Fertile from Nonfertile Males

Jerome H. Check, Joanne R. Liss

The UMDNJ, Robert Wood Johnson Medical School at Camden,
Cooper Hospital/University Medical Center, Department Ob/Gyn,
Division of Reproductive Endocrinology and Infertility, Camden, N.J., USA

In a previous study 135 infertile couples were evaluated to determine the pregnancy rate in 6 cycles. Included were all females who had a female factor(s) identified and thought to be fully corrected. Although a semen analysis was performed, no therapy was offered to the male partner for this time period. During the 6 months, males with normal motile densities (MD) according to the World Health Organization (WHO) standards (minimum of 10×10^6/ml motile sperm with $> 50\%$ normal morphology) showed a pregnancy rate of 83% as compared to 69% in those couples with semen specimens below WHO standards. The differences, however, were not statistically significant. These data make unreliable previous statistics which suggest the percentage of male factor as a cause of infertility in this population. They were based solely on determining what percentage of men in infertile couples have motile sperm counts below WHO standards. It had already assumed this to be the cause of the infertility. The data also provided serious questions as to the validity of performing semen analyses on patients and predicting male factor problems based on the results with the exception, perhaps, of the extremely poor motile densities.

We therefore considered the possibility that in couples who failed to conceive after 8 cycles despite all female factors having been corrected, males who exhibited subnormal semen parameters might be those with a greater likelihood of not achieving a pregnancy during the next 8 cycles. When a couple has had 8 cycles to achieve a pregnancy without success, it has been the experience of many that continuing on the same therapy will result in only a very small percentage of pregnancies. Theoretically the failure in these unexplained cases might be related to either an occult male or female factor problem, or possibly both. A priori one would think that those with subnormal semen parameters (despite the fact that these levels do not predict infertility during the first eight cycles of exposure after the correction of female factors) might certainly have problems with male fertility potential whereas, in men with levels above the minimum requirements by WHO standards, the cause of infertility may be related more to an occult female factor problem.

To test the hypothesis that if the semen analysis is not predictive of infertility during the first 6–8 months, this would at least predict the cause of infertility in those who failed during that time, we initiated a study using a donor probe. It was reasoned that men with a subnormal spermiogram would, in all probability, respond with a high pregnancy rate following therapeutic donor insemination (TDI) whereas those with normal semen analyses would demonstrate a much lower pregnancy rate following TDI. We report, herein, the pregnancy rates following TDI in couples who failed to conceive after 8 cycles when all known female factors had been corrected. All couples were separated into two groups, one in which the male partner had subnormal spermiograms and one group with normal spermiograms. Further evaluation included 70 additional couples who also failed to achieve pregnancies after 8 corrected cycles in which all males had normal semen parameters but where the hamster ova penetration test was either normal or subnormal, and the response to TDI also evaluated.

Materials and Methods

In study one, 88 couples who had a minimum of 12 months of infertility were selected. The requirement was that at least one infertility factor be identified and corrected in the female partner. Luteal phase defects were evaluated by late luteal phase timed endometrial biopsies in two consecutive cycles and dated according to the criteria of Noyes et al. [2]. Ovulation was treated by either clomiphene citrate, human menopausal gonadotropins or bromocriptine for hyperprolactinemia. Adequacy of therapy was evaluated by demonstrating a mature follicle by sonography [3] and also an adequate level of estradiol. The postcoital test was evaluated after at least 10 h after intercourse; an abnormality was considered if there were less than 3 sperm per high-power field with good linear progressive motion. The cervical factor was treated with either guaifenesin [4] or high dose estrogen and human menopausal gonadotropins (hMG) [5, 6]. Women who had laparoscopies and were diagnosed with endometriosis were included only if it was felt that after fulguration or vaporization there was no disease remaining. Those having tubal occlusion or adhesive disease were only to be included if it was felt that the problem was corrected following surgery.

Semen analyses were performed using a Makler chamber. Men with extremely low motile densities $< 2.5 \times 10^6$/ml were eliminated from the study. TDI was performed at mid-cycle in the female partner with the timing based on follicular maturation studies using ultrasound, estradiol, and progesterone criteria. The pregnancy rates after 8 cycles were then determined in groups that appeared to have normal semen analyses during the previous two baseline specimens evaluated versus those with subnormal semen specimens.

In study two, 70 men with unexplained infertility but semen analysis in two consecutive measurements also considered normal were treated with TDI before considering in vitro fertilization. These men had hamster ova penetration test performed with a technique described by Yanagimachi et al. [7]. Similarly, the pregnancy rate was determined in the female partner of those males having normal hamster tests versus those who fell below the level of 10% of the eggs penetrated. Two inseminations per cycle were performed in each female.

Results

The pregnancy results of the couples who failed to achieve a pregnancy after 8 months despite correction all female factors and subsequent treatment by TDI can be seen in table 1. The data do not support the initial hypothesis that a higher pregnancy rate is expected in men who had the lowest motile densities (during the initial 8 months). In fact, there were no statistical differences between the groups.

The data, however, do show that the motile density was not particularly predictive of a normal male in the sense that if this were a good indicator, we would not have found a high pregnancy rate in the group of failures during the first 8 months. There was, however, a respectable rate in both the subnormal and the normal males of 65 and 62%, respectively. Table 2 demonstrated that poor hamster ova penetration tests were not better than lower motile densities at predicting which couple would achieve a higher pregnancy rate following TDI.

In a group of 200 patients selected following the correction of all factors after 8 cycles and offered TDI but refused, there were 142 men considered as having a normal semen analysis and 58 men with a substandard analysis. A total of 34 patients achieved a pregnancy during the 8 months (17%) following continuation of the same therapy that had corrected all

Table 1. Pregnancy rates following therapeutic donor insemination correlated with the motile densities of the male partner in couples failing to conceive after 8 cycles of all female factors corrected

	Motile densities, millions/ml	
	WHO standards ($< 10 \times 10^6$/ml)	normal WHO standards ($\geq 10 \times 10^6$/ml)
Number of couples	15	73
Number pregnant	13	53
% pregnant	86.6	72.6

Table 2. Pregnancy rates during 8 cycles in couples conceiving after 8 cycles of all factors corrected according to whether the hamster ova penetration test was normal or subnormal (all men with normal motile densities)

	Men with subnormal hamster test	Men with normal hamster test
Number of couples	20	50
Number pregnant	13	31
% pregnant	65	62

the factors during the first 8 months. Fisher's exact test revealed no significant differences comparing TDI pregnancy results among the groups with normal or poor hamster ova penetration tests (p = 0.51) or comparing the pregnancy results from TDI in couples with previous low versus normal motile densities (p = 0.21).

Discussion

It has been estimated that the male contributes to infertility in 40–50% of all cases; these figures, however, are based on WHO standards. The data presented herein do not challenge the fact that the male is a cause of infertility even when the motile density is not extremely high. What this study does challenge is the fact that the male can be identified by the standard semen analysis. By demonstrating a high pregnancy rate following TDI in men whose semen specimens were subnormal vs. normal, whether it be motile density or the hamster ova penetration test, the data confirms that methods of evaluating male fertility potential cannot even predict the subfertile male who has failed to achieve a pregnancy after 8 cycles, even with all the female problems corrected. Thus, the results of this study underscore the need to find other tests that will better help us identify subfertile men. This study and its findings further emphasizes the extreme importance of identifying and correcting all female factors even in couples where the males who are thought to be subnormal since they may still achieve a high pregnancy rate after aggressive therapy of the female partner.

References

1   Check JH, Epstein R, Nowroozi K, Shanis BS, Wu CH, Bollendorf A: The hypo-osmotic swelling test as a useful adjunct to the semen analysis to predict fertility potential. Fertil Steril 1989;52:159–161.
2   Noyes RW, Hertig A, Rock J: Dating the endometrial biopsy. Fertil Steril 1950;1: 3–25.
3   Check JH, Goldberg BB, Kurtz A, Adelson HG, Rankin A: Pelvic sonography to help determine the appropriate therapy for luteal phase defects. Int J Fertil 1984;29: 156–158.
4   Check JH, Adelson HG, Wu CH: Improvement of cervical factor with guaifenesin. Fertil Steril 1982;37:707–708.
5   Check JH, Wu CH, Dietterich C, Lauer C, Liss J: The treatment of cervical factor with ethinyl estradiol and human menopausal gonadotropins. Int J Fertil 1986;31: 148–152.
6   Check JH: Treatment of cervical factor with combined high-dose estrogen and human menopausal gonadotropins. Fertil Steril 1980;33:562–563.
7   Yanagimachi R, Yanagimachi H, Rogers BJ: The use of zona-free animal ova as a test-system for the assessment of the fertilizing capacity of human spermatozoa. Biol Reprod 1979;15:471–476.

Jerome H. Check, MD, 7447 Old York Road, Melrose Park, PA 19126 (USA)

Colpi GM, Pozza D (eds): Diagnosing Male Infertility.
Prog Reprod Biol Med. Basel, Karger, 1992, vol 15, pp 237–239

Impact of in vitro Fertilization on Obstructive Azoospermia

A Preliminary Report

G. Schultes, G. Lunglmayr, E. Müller-Tyl, J. Huber

Department of Urology, Mistelbach General Infirmary and
First and Second Departments of Obstetrics and Gynaecology,
University of Vienna Medical School, Vienna, Austria

Congenital absence of the vas deferens is commonly found in males presenting with obstructive azoospermia. Treatment of this condition is still in an experimental stage. The development of alloplastic reservoirs gave rise to the idea to collect epididymal sperm for subsequent homologous insemination. However, results have not been encouraging so far.

Another interesting approach to treat congenital vas deferens aplasia has been reported by Silber and colleagues in 1988. Microsurgical aspiration of sperm from the proximal segment of the epididymis combined with IVF resulted in two pregnancies. We started a continuing trial in Austria to further assess this technology for management of obstructive azoospermia and congenital vas deferens aplasia, respectively. This is a preliminary report on 3 couples, who underwent epididymal sperm aspiration and IVF.

Three couples reported sterility in marriage over 2–9 years. There was no evidence of female factors. The women had regular menstrual cycles and ovulation and normal luteal phase was confirmed by ultrasound. An inconspicuous situs was found in the small pelvis on laparoscopy. Both fallopian tubes exhibited normal passage of the blue dye. All investigated hormone levels were within the normal range.

All males were azoospermic. On palpation the testicles revealed normal volume and serum concentrations of FSH were found within the normal range. On rectal examination the prostate was normal. There was no evidence of chronic inflammation. Testis biopsies were performed and normal spermatogenesis was found in semithin sections.

In 2 individuals epididymal pathology consisted of congenital vas deferens aplasia. The epididymis appeared slightly enlarged on both sides and the vas was not palpable. The third patient underwent bilateral vasectomy 15 years ago. In 1986 he got married again and requested refertilization. Bilateral microsurgical vaso-vasostomy was performed resulting in persisting azoospermia. In 1988 reassessment of the patient gave conclusive evidence of secondary intraepididymal block and subsequent bilateral vaso-epididymostomy was carried out without any success. The patient was referred to us for epididymal sperm aspiration in 1989.

According to our treatment protocol the women were prepared for IVF using clomiphene citrate/HMG stimulation. Oozyte development was monitored by ultrasound from cycle day 9 and 10,000 IU of HCG were injected 35 h prior to expected oozyte retrieval.

Prior to aspiration of oozytes the males underwent scrotal exploration under general or local anesthesia. Scrotal contents were extruded through a small incision and the epididymis was exposed. Under 10–40 × magnification with an operating microscope, a tiny incision was made with microscissors into the epididymal tunic to expose the tubule.

Beginning in the most distal part of the epididymis the first sample of epididymal fluid was immediately diluted in Menozo B2 medium and investigated for sperm and progressive motility. If motile sperm were visible fluid aspiration was continued over a period of at least 20 min to achieve maximal sperm concentration.

Great care was taken not to contaminate the specimen with blood. Careful hemostasis was achieved with microbipolar forceps. For continuing aspiration of epididymal fluid a commercially produced micro pipetter was used which allowed constant aspiration via capillary flow and capillary attraction.

Analysis of total sperm count and progressive motility in the aspirates was productive of the following results:

The patients presenting with congenital vas deferens aplasia yielded a total sperm count of 7.2 and 27.1 millions/ml, respectively. In the patient who presented with vaso-vasostomy and vaso-vasostomy failure a total sperm count of 4.0 millions/ml was achieved. Motility ranged between 5 and 20% immediately following dilution of aspirate in Menezo B2 medium.

A swim-up was performed on the specimens and the resulting fraction of sperm recovered had a motility between 17 and 36% with a progression of 6–25%.

Normal forms ranged between 12 and 18% after a wash and swim-up procedure. After sperm analysis oozyte retrieval was performed via transvaginal probe under ultrasound control. Between 2 and 7 oozytes were cultured in a humidified atmosphere of 5% CO_2 and air at 37 °C and incubated with sperm over 15–18 h. In two couples the oozytes were fertilized with epididymal sperm: in 1 patient with congenital vas deferens aplasia and in the other following epididymal sperm retrieval after vaso-epididymostomy failure.

Embryo transfer was carried out in 2 cases at the 4- and 8-cell stages of development and 1 patient conceived according to increasing concentrations of serum beta-HCG. It was the couple presenting with vaso-epididymostomy failure. Unfortunately, abortion occurred.

Our preliminary results confirm that fertilization of oozytes with sperm aspirated from proximal segments of the epididymis is possible in cases with congenital vas deferens aplasia or failed vasectomy reversal.

Dr. G. Schultes, 1st Department of Obstetrics and Gynecology,
University of Vienna Medical School, Spitalgasse 23, A–1090 Vienna (Austria)

Colpi GM, Pozza D (eds): Diagnosing Male Infertility.
Prog Reprod Biol Med. Basel, Karger, 1992, vol 15, pp 240–245

Should We Consider Ethical Limits in Diagnosing Male Infertility?

Antonio G. Spagnolo, Maria L. Di Pietro, Elio Sgreccia

Centre for Bioethics, The Sacred Heart Catholic University, School of Medicine, Rome, Italy

The desire on the part of the spouses for a child expresses the vocation to fatherhood and motherhood inscribed in conjugal love. Therefore, the suffering of spouses who cannot have children is a suffering that everyone must understand and properly evaluate [1]. According to the data in the literature [2], infertility affects about 15% of the couples who do not achieve pregnancy within 1 year of unprotected intercourse. One-third of this infertility is due to the male factors. Therefore, where a cure is possible the specialists should intervene with all the available resources.

However, the desire for a pregnancy and a child at all costs can often put the respect of the dignity of the person in second place. The aim of this paper is to highlight those aspects of the diagnostic and therapeutic techniques in male infertility that seem to threaten the respect for the patient and the early stages of human life.

Our considerations include: (a) the modalities of sperm collection; (b) sperm testing; (c) genetic, hormonal and instrumental tests, and (d) the training and education of specialists. These considerations do not include the ethical problems connected to the use of diagnostic data for therapy, for example: artificial insemination, homologous and heterologous in vitro fertilization, because they have been fully examined in other papers.

Sperm Collection

The analysis of the sperm is carried out with reference to the evaluation of some parameters (quantitative, morphological, immunity, etc.), so sperm collection is obviously an essential step in the diagnostic protocol of male infertility. There are no ethical problems in the examination of the sperm, but rather in the modality of sperm collection, which is usually carried out through masturbation. This is not only for moral reasons but also because of the psychological aspect of masturbation. As far as the Catholic moral is concerned [3], for example, masturbation is an intrinsically and seriously disordered act, which objectively constitutes the deliberate use of the sexual faculty outside normal conjugal relations and contradicts the finality of the faculty. However, some Catholic theologians [4] consider that at a subjective level masturbation for the purpose of sperm testing 'gives support to the reinterpretation of the morality of masturbation'. Nevertheless, the physicians and the patients who wish to respect the Catholic moral indications should not have to act against their conscience or their sensitiveness even if just to produce a sample for sperm analysis. This is why many patients experience a sense of discomfort, if only psychological, when asked to collect sperm through masturbation and they often refuse to do so, thus interrupting the diagnostic protocol. The alternative hypotheses proposed [5] so far (postcoital collection, collection of residual sperm from the urethra, squeezing the prostate gland and the seminal vesicles, etc.) are not valid from a scientific point of view so they cannot be proposed from an ethical point of view either. Some authors had sperm collected during intercourse [6] using a condom-shaped silastic device which resulted in the collection of significantly better quality semen than that produced by masturbation. Nevertheless, there is some doubt as to the scientific and moral validity of this technique. In order to be able to indicate a morally and psychologically valid technique, our center together with specialists in this field, is carrying out a study protocol for the evaluation of the possibility of using vibrating instruments which induce ejaculation [7] without erotic stimulation. In conclusion, scientists have to consider the modalities of sperm collection that will be proposed to the patient and these modalities should be both scientifically valid and should also respect the dignity of the person.

Sperm Testing

Once collected, the sperm is tested thoroughly. Besides the evaluation of, for example, the presence of certain substances (antibodies, adenosine triphosphate, etc.) or the motility and other morphological characteristics, some tests cannot be used because they present ethical problems. In particular, we refer to the tests that include the sperm-egg interaction, egg penetration and in vitro fertilization (IVF) [8]. Before the introduction of human IVF these tests used to be performed by evaluating the gamete interaction between the sperm and the acellular zona pellucida, derived from previous ovarian biopsies. This technique did not seem to present any ethical problems. This test was later carried out on animal oocytes (hamster egg test), thus creating interspecies fertilization which, since it seriously offends the dignity and nature of the human person, is ethically unacceptable. It is possible to justify this test because it is stopped before the fusion of the two genomes so that no zygote is formed. This argument is founded on an assumption that we do not agree with, that individual life does not begin at the moment of the penetration of the oocyte by the sperm, but rather with the formation of the zygote, after the fusion of the genomes. On the other hand, this fusion does seem to not occur between gametes of different species and therefore the process would only be limited to the pronuclear stage. There still remains the doubt as to whether the above-mentioned procedure is ethically acceptable. However, the test is definitely ethically unacceptable when human oocytes are used. In a previous document [9] our center states that as soon as the oocyte and the spermatozoon interact a *new system* immediately begins. This is confirmed by the beginning of a chain of activity [10] in which the two systems (oocyte and sperm) no longer operate independently from each other but the new system begins to operate like an individual unit. Immediately after the penetration of the oocyte, and as a consequence of this event, a whole series of biochemical and genetic modifications occur, leading to the development of the human individual which begins its own vital cycle from that moment.

These biological conclusions are the basis of some international documents [11] which invite governments to prohibit any in vitro creation of embryos with no hypothesis of a subsequent transfer into the uterus. So even if the procedure is stopped at the pronuclear stage it is ethically unacceptable. This is a case of real exploitation of embryos which live for a few hours and then are destroyed, the only aim being to obtain the prediction of the outcome of assisted reproduction. Nevertheless, some reports [8] have criticized the results of IVF as not measuring the fertilizing potential of a couple so this scientific reason no longer exists (as though there were a scientific reason justifying the exploitation of embryos!).

Genetic, Hormonal and Instrumental Tests

Other examinations can be requested for the diagnostic protocol for male infertility. We believe that some ethical conditions have to be respected: (a) after being fully informed and having understood the implications, the patient should be able to choose whether or not to perform the examination; (b) the specialist has to find the most suitable way of informing the patient of the results with respect to his pathology and psychological situation; (c) confidentiality has to be guaranteed but a dialogue between the spouses should be encouraged and the patient should be invited to inform his wife of the result; (d) a precise diagnostic indication is necessary since the examinations are very expensive and are not always cost effective. This last requirement is particularly important for the instrumental examinations (cavernosography, vesiculography, spermatic phlebography, testicular biopsy, etc.) which present a risk for the patient that can only be accepted if the patient is fully informed and if the risk/benefit ratio is in proportion.

Training and Education of Specialists

The specialists in this field ought to have some qualities which also have ethical value: professional competence, awareness of values, coherent clinical practice, willingness to collaborate. Professional competence is a priority requirement of the medical profession and includes the complexity and totality of medical knowledge. Since this competence is becoming more and more specialized the physician and biologist have to make an effort to continuously update their knowledge.

The awareness of values is in relation to the basic anthropology which inspires the physician: he cannot ignore that in his profession human values, including the value of health and life itself, are always involved. The richer the physician's awareness of these values, the more careful and sensitive his conscience will be.

Ethics is not only a theoretical science but when applied to medicine it has practical implications which highlight the values. It is important to point out that if this practical aspect is coherent with professional competence and an awareness of the values, the action itself becomes ethical, and the physician, the patient and the community all benefit.

Willingness to collaborate is fundamental in the physician-patient relationship and should also include all the people who are involved in the diagnostic and therapeutic process. Only by wanting to collaborate is it possible to prevent the fragmentation of the medical practice so that it does not lose its effectiveness and human significance.

Conclusions

This paper simply aims to underline some scientific and ethical aspects and implications of the diagnosis of male infertility and demonstrates that moral values do not represent an obstacle for or conflict with scientific research. Indeed, in fully respecting the dignity of the person and of procreation, some scientists have achieved results which previously seemed unattainable.

Moreover, the Vatican's Congregation of the Doctrine of the Faith's statement on respect for human life in its origin and the dignity of procreation is extremely significative [1]: scientists 'are to be encouraged to continue their research with the aim of preventing the causes of sterility and of being able to remedy them so that sterile couples will be able to procreate in full respect for their own personal dignity and that of the child to be born.'

References

1   Congregation for the Doctrine of the Faith: Instruction on Respect for Human Life in Its Origin and on the Dignity of Procreation. Vatican City, Vatican Polyglot Press, 1987, 2nd part, No. 8.
2   Comhaire FH, Farley TMM, Rowe PJ: The infertile couple: Definitions and standards; in Negro-Vilar A, et al (eds): Andrology and Human Reproduction. New York, Raven Press, 1988 pp 191–201.
3   Sacred Congregation for the Doctrine of the Faith: Declaration on Certain Questions Concerning Sexual Ethics. Vatican City, Vatican Polyglot Press, 1975, No 9.
4   Keane, PS: Sexual Morality. A Catholic Perspective. Dublin, Gill & Macmillan, 1980 p 69.
5   Dunn HP: Semen examination. Linacre Quarterly 1987;54,1:88–91.
6   Zavos PM: Characteristics of human ejaculates collected via masturbation and a new Silastic seminal fluid collection device. Fertil Steril 1985;43:491.
7   Brindley GS: Reflex ejaculation under vibratory stimulation in paraplegic men. Paraplegia 1981;19:299.

8   Cohen J: Bioassays for male infertility; in Negro-Vilar A, et al (eds): Andrology and
     Human Reproduction. New York, Raven Press, 1988, pp 17–23.
9   The Sacred Heart Catholic University's Centre for Bioethics: Identity and status of
     the human embryo. Med Morale 1989;4(suppl):15–25.
10   Wassarman PM: The biology and chemistry of fertilization. Science 1987;235:553–
     560.
11   Parliamentary Assembly of The Council of Europe: Recommendations 1046 (1986)
     and 1100 (1989) on the use of human embryos and foetuses.

Dr. Antonio G. Spagnolo, Centro di Bioetica, Università Cattolica S. Cuore,
Largo Francesco Vito, 1, I–00168 Roma (Italy)

Subject Index